Cambridge Tracts in Mathematics
and Mathematical Physics

No. 30

THE DISTRIBUTION OF PRIME NUMBERS

THE DISTRIBUTION OF PRIME NUMBERS

BY

A. E. INGHAM, M.A.

Fellow of King's College, Cambridge
Sometime Fellow of Trinity College, Cambridge

CAMBRIDGE UNIVERSITY PRESS
Cambridge, New York, Melbourne, Madrid, Cape Town,
Singapore, São Paulo, Delhi, Tokyo, Mexico City

Cambridge University Press
The Edinburgh Building, Cambridge CB2 8RU, UK

Published in the United States of America by
Cambridge University Press, New York

www.cambridge.org
Information on this title: www.cambridge.org/9780521397896

First published 1932
Reprinted by Stechert-Hafner Service Agency 1964
Reissued as a paperback with a Foreword by R. C. Vaughan
in the Cambridge Mathematical Library series 1990
Reprinted 1992, 1995

A catalogue record for this publication is available from the British Library

Library of Congress Cataloguing in Publication data

ISBN 978-0-521-39789-6 Paperback

CONTENTS

FOREWORD

As an introduction to the distribution of primes Ingham's tract is still unsurpassed, combining an economy of detail with a clarity of exposition which eases the novice's way into this, sometimes technically ferocious, area. In spite of the fact that it has been out of print for many years I usually place it at, or near, the top of the reading list for graduate students.

When Ingham wrote his tract, the theory of the distribution of primes depended almost exclusively on the theory of the Riemann zeta-function. To a large extent this is still true, but there have been a number of important developments which depend, at least in part, on various aspects of sieve theory or the use of exponential sums of various kinds.

In 1934 the theory of the distribution of primes in an arithmetic progression $qn + a$ with $(q, a) = 1$ was in all essentials the precise analogue of the case $q = 1$ and so was ignored by Ingham. In the intervening years there has been significant progress concerned particularly with the situation when the modulus q is large, and this means that a modern exposition of the theory of primes has to cover a good deal more ground. For a good introduction to this material see Davenport (1980).

Ingham was very careful to connect his exposition with the original research which led to the theory he describes. There have been many refinements and developments of this theory in the intervening years and below we indicate some of them as they relate to Ingham's text. Ingham's brief was to leave the deeper properties of the Riemann zeta-function for exposition by Titchmarsh in a parallel tract, Titchmarsh (1930), and much of the work on the distribution of primes over the last 50 years is intimately related to such properties of the zeta-function.

Introduction

§ 5. There have been a number of attempts to obtain more precise estimates for $\pi(x)$ by methods related to those of Chebyshev, and a good deal is now known about them and their limitations. See Nair (1982).

§ 6. Edwards (1974) has devoted a book to a study of Riemann's seminal memoir and all later work which stems from it, and is perhaps now the most accessible source for much of the early material.

§ 7. Landau's function theoretic method mentioned here has been central to the improved results described in the comments on Chapter III, § 13 below.

§ 8. Selberg (1949) discovered a 'real variable', or 'elementary' proof of the prime number theorem. Thus the view expressed in § 8 is nugatory. Selberg first establishes the approximate identity

$$\psi(x)(\log x)+\sum_{n\leq x}\Lambda(n)\psi(x/n)=2x(\log x)+O(x),$$

or rather an approximate identity which is essentially equivalent to this. That is relatively easy. Harder, and harder to motivate, is the Tauberian process of extracting asymptotic information about $\psi(x)$. Several of the numerous expositions are unsatisfactory in this regard. There is a variant of the proof in Erdős and Selberg (1949), and of the expositions Diamond (1982) and Hardy and Wright (1979, Chapter XXII) are reliable without getting bogged down in the details.

A number of authors have adapted Selberg's method to obtain quite a good error term in the prime number theorem. The best estimate so far obtained by this method is

$$O(x\exp(-\log^a x))$$

with a any number smaller than 1/6. This is due to Lavrik and Sobirov (1973) and should be compared with Theorem 23 and the improved result stated below.

For a completely different elementary proof, see Daboussi (1984).

§ 9. The truth, or otherwise, of the Riemann hypothesis has still not been established, despite quite frequent claims which are often sensationalised by the media.

§ 10. Electronic computers have enabled the calculations to be greatly extended. See, for example, Riesel (1985), Chapters 1 and 2. A small correction has been made to the table on page 7. Skewes (1933, 1955) showed that, contrary to everyone's belief in the 1920s and early 1930s, Littlewood's theorem is not a pure 'existence' theorem, and that the first sign change occurs for some x not exceeding $10_4(3)$ where

$$10_1(x)=10^x, 10_2(x)=10^{10_1(x)},$$

and so on. See pages 110–12 of Littlewood's delightful *Miscellany*, (Littlewood, 1986) for an account of this work. Later, by a different method, Lehman (1966) reduced Skewe's constant considerably, to $1{\cdot}65\times10^{1165}$, and recently te Riele (1986) has shown that between $6{\cdot}62\times10^{370}$ and $6{\cdot}69\times10^{370}$ there are 10^{180} successive integers x for which $\pi(x)>\mathrm{li}(x)$.

§ 11. Titchmarsh's tract was later expanded into a celebrated text (Titchmarsh, 1986) which is still the standard work on the subject. With regard to upper bounds for $p_{n+1}-p_n$, Ingham himself (1937) made the most startling progress by showing that the bound he states here holds with $\vartheta>5/8$. Heilbronn (1933) had earlier obtained the lower bound $1-1/250$ and more recently Montgomery (1969) (see also Montgomery, 1971), Huxley (1972), Iwaniec and Jutila (1979), Heath-Brown and Iwaniec (1979), Iwaniec and Pintz (1984) and Mozzochi (1986), have obtained 3/5, 7/12, 13/23, 11/20, 23/42 and 1051/1921, respectively. In fact there has been much activity in this area over the last 20 years. For an account of some of this see Chapter 12 of Ivić (1985).

It is now known that $p_{n+1}-p_n$ is sometimes significantly larger than $\log p_n$. Following work of Erdős (1935), Rankin (1938, 1963) and Schönhage (1963), Maier and Pommerance (1990) have shown that

$$\limsup_{n\to\infty}\frac{(p_{n+1}-p_n)(\log\log\log p_n)^2}{(\log p_n)(\log\log p_n)(\log\log\log\log p_n)}\geqslant 4\mathrm{e}^{\gamma}/c,$$

where γ denotes Euler's constant and c satisfies $c=3+\mathrm{e}^{-c}$.

In the opposite direction it is still not known whether

$$\liminf_{n\to\infty}(p_{n+1}-p_n)<\infty.$$

However, Erdős (1940), established that there is a real number $C<1$ such that

$$\liminf_{n\to\infty}\frac{p_{n+1}-p_n}{\log p_n}\leqslant C.$$

Bombieri and Davenport (1966) showed that $C=0{\cdot}467$ is permissible, and after some small improvements by Pil'tai (1972) and Huxley (1973, 1977), Maier (1988) has obtained $C=0{\cdot}248$.

In much of the work in connection with differences between consecutive primes, sieve theory has played a significant role. The two standard works on sieves are Halberstam and Richert (1974) and Hooley (1976).

Chapter I

§ 3. For the further developments of sieve theory see the texts quoted above. The Goldbach binary problem is still unsolved. The constant k mentioned here has now been shown to satisfy $k \leqslant 19$ (Riesel and Vaughan, 1983), and Vinogradov (1937) has shown that every large odd number is the sum of three primes. Also Montgomery and Vaughan (1975) have shown that there is a positive constant c such that the number of even numbers not exceeding x which are not the sum of two primes is at most $O(x^{1-c})$, and Chen (1973, 1978) has shown that every large even number is the sum of a prime and a number having at most two prime factors.

Chapter II

§ 3. With regard to the reference in the footnote and more recent material see Titchmarsh (1986), Chapters V, VI, VIII, XIV.

§ 4. With regard to the comment on (10) at the end of this section, see the remarks below on § 9 of Chapter III.

§ 11. There are now other proofs of the prime number theorem which require only information about $\zeta(s)$ in the closed half-plane $\sigma \geqslant 1$, due to Wirsing (1967) and Halász (1969). They are important because they have arisen as special cases of solutions to quite general problems concerning mean values of multiplicative functions, and have spawned a large volume of investigations into mean value theorems for general classes of additive and multiplicative functions.

§ 12. In view of Selberg's elementary proof of the prime number theorem (Selberg, 1949) there is no longer any distinction between elementary and transcendental here.

Chapter III

§ 8. A modern reference for footnote 2 on page 58 is Titchmarsh (1986), (2.12.8).

§ 9. With a view to calculating a relatively large value for the constant a, there is an interesting extremal question with regard to non-negative cosine polynomials with non-negative coefficients akin to that mentioned at the end of § 4 of Chapter II. Here one is interested in the extremal value of

$$(\sqrt{c_1}-\sqrt{c_0})^2/(c_0+c_1+\cdots+c_m).$$

This question has not been answered completely. See Kondrat'ev (1977).

§ 11. In view of alternative proofs of the prime number theorem with reasonable error terms, and in particular Selberg's elementary method, one can ask for an inverse of Theorem 22, i.e. what zero-free region for the zeta-function can be deduced from a given error term in the prime number theorem. Turán (1950) discovered that an error term of the form

$$O(x\exp(-a(\log x)^b))$$

implies that $\zeta(s)$ is non-zero for

$$\sigma \geqslant 1-c(\log(2+|t|))^{(b-1)/b}$$

and in this special case this gives the desired converse. In particular this shows that a non-trivial zero-free region can be obtained by an 'elementary' proof.

Turán found many variants of the above theorem, and they are described in his book (Turán, 1984). Much of this work is intimately connected with research on irregularities of distribution, the topic treated by Ingham in Chapter V.

Ultimately, Pintz (1982a, b, 1984) established a converse to Theorem 22 which is valid under very general conditions.

§ 13. The zero-free region has been improved. It is now known that $\zeta(s)$ has no zeros in a domain of the form

$$\sigma > 1-a(\log t)^{-b}(\log\log t)^{-c}$$

with $b<1$. Chudakov (1936) was the first to obtain a region of this kind, with b any real number exceeding $7/8$ and $c=0$. The best value so far obtained for b is $2/3$ due to Korobov (1958) and Vinogradov (1958). With regard to the value of c, Korobov and Vinogradov apparently claim more ($c=0$) than can be established by their methods, namely $c=1/3$, and no retraction has ever been made. Ingham, in a review (1964) is justifiably scathing.

These improved zero-free regions are a consequence of estimates for exponential sums that follow from a mean value theorem for such sums known as the Vinogradov Mean Value Theorem. The exponential sum estimates imply bounds for the zeta-function in the neighbourhood of the line $\sigma=1$, and the zero-free region is obtained then by an application of the function theoretic method of Landau mentioned in § 7 of the Introduction.

The zero-free region mentioned above with $b=2/3$, $c=1/3$, when applied to Theorem 22, gives at once the best error term known in the prime number theorem, i.e. the error

$$O(x\exp(-(\log x)^{3/5}(\log\log x)^{-1/5})).$$

There are good expositions of this material in Ivić (1985) Chapters 6 and 12 and Richert (1967).

§ 14. For some discussions related to the Riemann hypothesis see Edwards (1974), § 1.9 and the notes to Chapter 1 of Ivić (1985), § 4.F of Ribenboim (1988) and Chapters XIV and XV of Titchmarsh (1986).

Chapter IV

§ 2. For developments on the distribution of zeros see Ivić (1985), § 1.4 and Chapter 11, and Titchmarsh (1986), Chapter 9.

§§ 5 and 6. The explicit formulae established here are the prototypes of many analogous formulae which have been discovered in other, not always closely related, areas. Apart from the generalisations by Weil (1952), and to Dirichlet L-functions, for which excellent sources are Chapter 5 of Patterson (1988) and Chapter 19 of Davenport (1980) respectively, there are, for instance, explicit formulae connected with rational functions over finite fields (see Chapter II of Schmidt (1976)) and with Riemann surfaces (see Selberg (1956), Hejhal (1976)).

§ 8. With regard to the order of the difference $\psi(x)-x$, see the comments below to Chapter V, § 8.

Chapter V

This chapter has become very dated, at least with regard to the proof of Littlewood's theorem, Theorem 34, largely as a consequence of Ingham's own investigations. In Ingham (1936), by considering a suitable weighted average of $\psi(x)-x$, he was led to the weighted partial sum

$$\sum_{0<\gamma\leq T}(1-\gamma/T)\frac{\sin\gamma t}{\gamma}$$

and was able thereby to avoid the use of the Phragmén–Lindelöf Theorem. Later, in the notes to Hardy's collected works (Hardy, 1967, page 99) he briefly indicates a further improvement. The later work of Skewes with regard to the first sign change of $\pi(x)-\mathrm{li}(x)$ mentioned above in the comments to § 10 of the Introduction is based on a variant of the proof given here. As it stands the proof is non-effective as far as the determination of the implicit constant is concerned. However, Skewes avoids this by a small change in the division of cases, in essence Riemann hypothesis 'almost true' and its negation. The later work of Lehman and te Riele is based on Turán's method, mentioned in the comments to § 11 of Chapter III. All work of this kind requires a quantitative theorem on diophantine approximation, such as Dirichlet's theorem, Theorem J, or something equivalent. Unfortunately Dirichlet's theorem is a theorem on homogeneous diophantine approximation whereas the underlying problem in this instance is an inhomogeneous one - ideally one wishes to arrange that, for suitable t, each of the quantities $\sin(\gamma t)$ with $0<\gamma\geq T$ are all close to $+1$ or all close to -1. However, the inhomogeneous theory (Kronecker's theorem - see Chapter XXIII of Hardy and Wright (1979)) only gives qualitative estimates. Turán's method overcomes this difficulty and provides quantitative estimates in an inhomogeneous situation. The estimates are relatively weak and do not improve on Littlewood's theorem. However, they are more flexible and more amenable to calculation. For further reading in this area see Chapter 12 of Ivić (1985) and, especially, Turán (1984) and Pintz (1982a, b, 1984).

Montgomery has suggested that, provided that the Riemann hypothesis holds and linear forms in the imaginary parts of the zeros above the real axis are not abnormally small, then

$$\limsup_{x\to\infty} (\psi(x)-x)x^{-1/2}(\log\log\log x)^{-2} = 1/(2\pi)$$

and

$$\liminf_{x\to\infty} (\psi(x)-x)x^{-1/2}(\log\log\log x)^{-2} = -1/(2\pi).$$

REFERENCES

Bombieri, E. and Davenport, H. (1966). Small differences between prime numbers, *Proc. Roy. Soc.*, A**293**, 1-18.

Chen, J. R. (1973, 1978). On the representation of a large even integer as the sum of a prime and the product of at most two primes, I, II, *Sci. Sinica*, **16**, 157-176 and **21**, 421-430.

Chudakov, N. G. (1936). On zeros of the function $\zeta(s)$, *C. R. Acad. Sci. URSS* **1**(x), 201-204.

Davenport, H. (1980). *Multiplicative Number Theory*, second edition, GTM 74, Springer-Verlag, Berlin.

Daboussi, H. (1984). Sur le théorème des nombres premiers, *C. R. Acad. Sci. Paris*, Sér. I, **298**, 161-164.

Diamond, H. (1982). Elementary methods in the study of the distribution of prime numbers, *Bull. Amer. Math. Soc.*, **7**, 553-589.

Edwards, H. (1974). *Riemann's Zeta Function*, Academic Press, New York.

Erdős, P. (1935). On the difference of consecutive primes, *Quart. J. Pure & Appl. Math., Oxford*, **6**, 205-213.

Erdős, P. and Selberg, A. (1949). On a new method in elementary number theory which leads to an elementary proof of the prime number theorem, *Proc. Nat. Acad. Sci.* **35**, 374-384.

Halász, G. (1968). Über die Mittelwert multiplikativer zahlentheoretischer Funktionen, *Acta Math. Acad. Sci. Hungar.* **19**, 365-403.

Halberstam, H. and Richert, H.-E. (1974). *Sieve Methods*, Academic Press, New York.

Hardy, G. H. (1967). *Collected Papers of G. H. Hardy*, including joint papers with J. E. Littlewood and others, edited by a committee appointed by the London Mathematical Society, vol II, Oxford, Clarendon Press.

Hardy, G. H. and Wright, E. M. (1979). *An Introduction to the Theory of Numbers*, Clarendon Press, Oxford.

Heath-Brown, D. R. and Iwaniec, H. (1979). On the difference between consecutive primes, *Bull. Amer. Math. Soc.* NS, **1**, 758-760.

Hejhal, D. (1976). *The Selberg Trace Formula for PSL*(2, ℝ), LNM 548, Springer-Verlag, Berlin.
Hooley, C. (1976). *Applications of Sieve Theory to the Theory of Numbers*, CT 70, Cambridge University Press.
Huxley, M. N. (1972). On the difference between consecutive primes, *Inventiones Math.*, **15**, 164–170.
Huxley, M. N. (1973). Small differences between consecutive primes I, *Mathematika*, **20**, 229–232.
Huxley, M. N. (1977). Small differences between consecutive primes II, *Mathematika*, **24**, 142–152.
Ingham, A. E. (1936). A note on the distribution of primes, *Acta Arithmetica*, **1**, 201–211.
Ingham, A. E. (1964). *Mathematical Reviews*, **28**, #3954.
Ivić, A. (1985). *The Riemann Zeta-Function*, John Wiley & Sons, New York.
Iwaniec, H. and Jutila, M. (1979). Primes in short intervals, *Arkiv f. Mat.*, **17**, 167–176.
Iwaniec, H. and Pintz, J. (1984). Primes in short intervals, *Monatsh. Math.*, **98**, 115–143.
Kondrat'ev, V. P. (1977). Some extremal properties of positive trigonometric polynomials, *Mat. Zametki*, **22**, 371–374.
Korobov, N. M. (1958). Estimates of trigonometric sums and their applications, *Usp. Mat. Nauk*, **13**, 185–192.
Lavrik, A. F. and Sobirov, A. Š. (1973). On the remainder term in the elementary proof of the prime number theorem, *Dokl. Akad. Nauk SSSR*, **211**, 534–536.
Lehman, R. S. (1966). On the difference $\pi(x) - \mathrm{li}(x)$, *Acta Arithmetica*, **11**, 397–410.
Littlewood, J. E. (1986). *Littlewood's Miscellany*, edited by B. Bollobás, Cambridge University Press.
Maier, H. (1988). Small differences between prime numbers, *Michigan Math. J.*, **35**, 323–344.
Maier, H. and Pommerance, C. (1990). Unusually large gaps between consecutive primes, *J. Am. Math. Soc.*, to appear.
Mozzochi, J. (1986). On the difference between consecutive primes, *J. Number Theory*, **24**, 181–187.
Montgomery, H. L. (1969). Zeros of *L*-functions, *Inventiones Math.*, **8**, 346–354.
Montgomery, H. L. (1971). *Topics in Multiplicative Number Theory*, LNM 227, Springer-Verlag, New York.
Montgomery, H. L. and Vaughan, R. C. (1975). The exceptional set in Goldbach's problem, *Acta Arithmetica*, **27**, 353–370.
Nair, M. (1982). On Chebyshev-type inequalities for primes, *Am. Math. Monthly*, **89**, 126–129.
Patterson, S. J. (1988). *An Introduction to the Theory of the Riemann Zeta Function*, Cambridge University Press.
Pil'tai, G. Z. (1972). On the size of the difference between consecutive primes, *Issled. teor. chisel*, 73–79.

Pintz, J. (1982a). Oscillatory properties of the remainder term of the prime number formula, in *Studies in Pure Mathematics, To the Memory of Paul Turán*, Birkhaüser Verlag, Basel, 551–560.

Pintz, J. (1982b). On the mean-value of the remainder term of the prime number formula, *Proc. Banach Centre Symposium in Analytic Number Theory*, Warsaw, Vol XVII.

Pintz, J. (1984). On the remainder term of the prime number formula and the zeros of Riemann's zeta-function, *Proc. Journées Arithmétiques* (Nordwijkerhout, Netherlands, 1983) LNM 1068, Springer-Verlag, Berlin, 186–197.

Rankin, R. A. (1938). The difference between consecutive prime numbers, *J. London Math. Soc.*, **13**, 242–247.

Rankin, R. A. (1963). The difference between consecutive prime numbers, V, *Proc. Edinburgh Math. Soc.*, (2) **13**, 331–332.

Ribenboim, P. (1988). *The Book of Prime Number Records*, Springer-Verlag, Berlin.

Richert, H.-E. (1967). Zur Abschätzung der Riemannschen Zetafunktion in der Nähe der Vertikalen $\sigma = 1$, *Math. Ann.*, **169**, 97–101.

te Riele, H. J. J. (1986). On the sign of the difference $\pi(x) - \mathrm{li}(x)$, Report NM_R8609, Centre for Math. and Comp. Science, Amsterdam, and *Math. Comp.*, **48**, 1987, 667–681.

Riesel, H. (1985). *Prime Numbers and Computer Methods for Factorization*, Birkhäuser Verlag, Basel.

Riesel, H. and Vaughan, R. C. (1983). On sums of primes, *Arkif för Matematik*, **21**, 45–74.

Schmidt, W. M. (1976). *Equations over Finite Fields. An Elementary Approach*, LNM 536, Springer-Verlag, Berlin.

Selberg, A. (1949). An elementary proof of the prime number theorem, *Ann. Math.*, **50**, 305–313.

Selberg, A. (1956). Harmonic analysis and discontinuous groups in weakly symmetric Riemannian spaces with applications to Dirichlet series, *J. Indian Math. Soc.*, **20**, 47–87.

Schönhage, A. (1963). Eine Bemerkung zur Konstruktion grosser Primzahllücken, *Arch. Math.*, **14**, 29–30.

Skewes, S. (1933). On the difference $\pi(x) - \mathrm{li}(x)$, *J. London Math. Soc.*, **8**, 277–283.

Skewes, S. (1955). On the difference $\pi(x) - \mathrm{li}(x)$, *Proc. London Math. Soc.*, **5**(3), 48–70.

Titchmarsh, E. C. (1930). *The Zeta-Function of Riemann*, CT 26, Cambridge University Press.

Titchmarsh, E. C. (1986). *The Theory of the Riemann Zeta-Function*, second edition, Clarendon Press, Oxford.

Turán, P. (1950). On the remainder term of the prime-number formula II, *Acta Math. Acad. Sci. Hung.*, **1**, 155–166.

Turán, P. (1984). *A New Method in the Analysis and its Applications*, Wiley-Interscience, New York.

Vinogradov, I. M. (1937). Some theorems concerning the theory of primes, *Mat. Sb.*, **2**(44), 179–195.

Vinogradov, I. M. (1958). A new estimate for $\zeta(1+it)$, *Izv. Akad. Nauk SSSR, Ser. Mat.*, **22**, 161–164.
Weil, A. (1952). Sur les 'formules explicites' de la théorie des nombres premiers, *Comm. Sém. Math. Univ. Lund, Tome Supplémentaire*, 252–265.
Wirsing, E. (1967). Das asymptotische Verhalten von Summen über multiplikative Funktionen II, *Acta Math. Acad. Sci. Hungar.*, **18**, 411–467.

R. C. Vaughan
Imperial College, London

PREFACE

The subject of this tract is the theory of the distribution of the prime numbers in the series of natural numbers. A chapter on the 'elementary' theory has been included for its historical interest and for the intrinsic interest of the methods employed, but the major part of the book is devoted to the analytical theory founded on the zeta-function of Riemann. The tract is thus a companion to No. 26 of the series, 'The zeta-function of Riemann' by Prof. E. C. Titchmarsh, published in 1930, but the logical sequence of the two volumes is the reverse of the chronological order of publication. The part of the theory of the zeta-function here required is what may be called the 'classical' theory, and comprises roughly those properties summarised by Prof. Titchmarsh in his Introduction. This is expounded in detail in the present volume, which is thus complete in itself (apart from a few isolated references to Titchmarsh which do not affect the understanding of the book as a whole); and the relevant parts may serve as an introduction to the more profound study of the zeta-function in the companion volume. The present tract is not intended exclusively for specialists, for whom the more comprehensive treatises of Landau, *Handbuch der Lehre von der Verteilung der Primzahlen* and *Vorlesungen über Zahlentheorie*, are already available; it aims rather at making the subject accessible to a wider circle of readers.

This volume like its companion has its origin in the Bohr-Littlewood manuscript referred to by Prof. Titchmarsh in his preface. This manuscript forms the basis of the present version, but a complete revision was found desirable in order to bring the work up-to-date and to take account of improvements of technique introduced since the preparation of the original. In the task of revision I derived much assistance from lecture notes kindly placed at my disposal by Prof. Littlewood. My indebtedness to the two books of Landau already referred to will be too obvious to readers of those works to need special emphasis here. The proof-sheets have been read by Prof. H. Bohr and Prof. J. E. Littlewood, the authors of the original manuscript, and also by Prof. G. H. Hardy, Dr A. Zygmund, Mr R. M. Gabriel, and Mr C. H. O'D. Alexander, and to these my thanks are due for a number of corrections and improvements. To Prof. N. Wiener I am indebted for some valuable comments on the concluding sections of Chapter II.

A. E. I.

THE DISTRIBUTION OF PRIME NUMBERS

INTRODUCTION

1. The positive integers other than 1 may be divided into two classes, prime numbers (such as 2, 3, 5, 7) which do not admit of resolution into smaller factors, and composite numbers (such as 4, 6, 8, 9) which do. The prime numbers derive their peculiar importance from the 'fundamental theorem of arithmetic' that a composite number can be expressed in one and only one way as a product of prime factors. A problem which presents itself at the very threshold of mathematics is the question of the distribution of the primes among the integers. Although the series of prime numbers exhibits great irregularities of detail, the general distribution is found to possess certain features of regularity which can be formulated in precise terms and made the subject of mathematical investigation.

We shall denote by $\pi(x)$ the number of primes not exceeding x; our problem then resolves itself into a study of the function $\pi(x)$. If we examine a table of prime numbers, we observe at once that, however extensive the table may be, the primes show no signs of coming to an end altogether, though they do appear to become on the average more widely spaced in the higher parts of the table. These observations suggest two theorems which may be taken as the starting-point of our subject. Stated in terms of $\pi(x)$, these are the theorems that $\pi(x)$ tends to infinity, and $\pi(x)/x$ to zero, as x tends to infinity.

2. The first theorem—that there exists an infinite number of primes—was proved by Euclid (*Elements*, Book 9, Prop. 20). In essentials his proof is as follows. Let P be a product of any finite set of primes, and let $Q = P + 1$. The integers P and Q can have no prime factor in common, since such a factor would divide $Q - P = 1$, which is impossible. But Q (being greater than 1) must be divisible by some prime. Hence there exists at least one prime distinct from those occurring in P. If there were only a finite number of primes altogether, we could take P to be the product of all primes, and a contradiction would result. The argument really gives a little more. It shows that, if p_n is the nth prime (so that $p_1 = 2$, $p_2 = 3$, $p_3 = 5$, ...), the integer $Q_n = p_1 p_2 \dots p_n + 1$ is divisible by some p_m with $m > n$, so that $p_{n+1} \leqslant p_m \leqslant Q_n$; from which we may infer, by induction, that

$$(1) \qquad p_n < 2^{2^n}.$$

3. In 1737 Euler proved the existence of an infinity of primes by a new method, which shows moreover that

$$(2) \qquad \text{the series } \sum_{n=1}^{\infty} \frac{1}{p_n} \text{ is divergent.}$$

Euler's work is based on the idea of using an identity in which the primes appear on one side but not on the other. Stated formally his identity is

$$(3) \qquad \sum_{n=1}^{\infty} n^{-s} = \prod_p (1 + p^{-s} + p^{-2s} + \dots) = \prod_p (1 - p^{-s})^{-1},$$

where the products are over all primes p. Euler's contribution to the subject is of fundamental importance; for his identity, which may be regarded as an analytical equivalent of the fundamental theorem of arithmetic, forms the basis of nearly all subsequent work.

The theorems (1) and (2) resemble one another in that they each add something (though in different ways) to the statement that the number of primes is infinite.

4. The question of the diminishing frequency of primes was the subject of much speculation before any definite results emerged. The problem assumed a much more precise form with the publication by Legendre in 1808 (after a less definite statement in 1798) of a remarkable empirical formula for the approximate representation of $\pi(x)$. Legendre asserted that, for large values of x, $\pi(x)$ is approximately equal to

$$(4) \qquad \frac{x}{\log x - B},$$

where $\log x$ is the natural (Napierian) logarithm of x and B a certain numerical constant[1]—a theorem described by Abel (in a letter written in 1823) as the 'most remarkable in the whole of mathematics'. A similar, though not identical, formula was proposed independently by Gauss. Gauss's method, which consisted in counting the primes in blocks of a thousand consecutive integers, suggested the function $1/\log x$ as an approximation to the average density of distribution ('number of primes per unit interval') in the neighbourhood of a large number x, and thus

$$\int_2^x \frac{du}{\log u} \tag{5}$$

as an approximation to $\pi(x)$. Gauss's observations were communicated to Encke in 1849, and first published in 1863; but they appear to have commenced as early as 1791 when Gauss was fourteen years old.[2] In the interval the relevance of the function (5) was recognised independently by other writers.[3] For convenience of notation it is usual to replace this function by the 'logarithmic integral'

$$\operatorname{li} x = \lim_{\eta \to +0} \left(\int_0^{1-\eta} + \int_{1+\eta}^x \right) \frac{du}{\log u},$$

from which it differs only by the constant $\operatorname{li} 2 = 1{\cdot}04\ldots$.

The precise degree of approximation claimed by Gauss and Legendre for their empirical formulae outside the range of the tables used in their construction is not made very explicit by either author, but we may take it that they intended to imply at any rate the 'asymptotic equivalence' of $\pi(x)$ and the approximating function $f(x)$, that is to say that $\pi(x)/f(x)$ tends to the limit 1 as x tends to infinity. The two theorems which thus arise, corresponding to the two forms of $f(x)$, are easily shown to be equivalent to one another and to the simpler relation

$$\frac{\pi(x)}{\dfrac{x}{\log x}} \to 1 \text{ as } x \to \infty; \tag{6}$$

but the distinction between (4) and (5), and the value of B in (4),

[1] Legendre **1a**, 19; **1b**, 394; **2**, ii, 65. The references in heavy type are to the bibliography at the end of the tract.

[2] Gauss **1**, ii, 444–447; x_1, 11.

[3] Dirichlet, *Werke*, i, 372, footnote **; Chebyshev **1**, **2**; Hargreave **1**, **2**.

become important if we enquire more closely into the order of magnitude of the 'error' $\pi(x) - f(x)$. The proposition (6), which is now known as the 'prime number theorem', is the central theorem in the theory of the distribution of primes. The problem of deciding its truth or falsehood engaged the attention of mathematicians for about a hundred years.

5. The first theoretical results connecting $\pi(x)$ with $x/\log x$ are due to Chebyshev. In 1848 he showed (among other things) that, if the ratio on the left of (6) tends to a limit at all, the limit must be 1; and in 1850 that this ratio lies between two positive constants a and A for all sufficiently large values of x, so that the function $x/\log x$ does at any rate represent the true order of magnitude of $\pi(x)$. These results constituted an advance of the first importance, but (as Chebyshev himself was well aware) they failed to establish the essential point, namely the existence of $\lim \pi(x)/(x/\log x)$. And, although the numerical bounds (a, A) obtained by Chebyshev were successively narrowed by later writers (particularly Sylvester), it came to be recognised in due course that the methods employed by these authors were not likely to lead to a final solution of the problem.

6. The new ideas which were to supply the key to the solution were introduced by Riemann in 1859, in a memoir which has become famous, not only for its bearing on the theory of primes, but also for its influence on the development of the general theory of functions. Euler's identity had been used by Euler himself with a fixed value of s ($s = 1$), and by Chebyshev with s as a real variable. Riemann now introduced the idea of treating s as a complex variable and studying the series on the left of (3) by the methods of the theory of analytic functions. This series converges only in a restricted portion of the plane of the complex variable s, but defines by continuation a single-valued analytic function regular at all finite points except for a simple pole at $s = 1$. This function is called the 'zeta-function of Riemann', after the notation $\zeta(s)$ adopted by its author.

Although Riemann is not primarily concerned with approximations to $\pi(x)$, his analysis shows clearly that this function is intimately bound up with the properties of $\zeta(s)$, and in particular with the distribution of its zeros in the s-plane. Riemann enunciated a number of important theorems concerning the zeta-function, together with a remarkable identity connecting $\pi(x)$ with its zeros, but he gave in most cases only insufficient indications of proofs. The problems raised by Riemann's memoir inspired in due course the fundamental researches of Hadamard in the theory of integral functions, the results of which at last removed some of the obstacles which for more than thirty years had barred the way to rigorous proofs of Riemann's theorems. The proofs sketched by Riemann were completed (in essentials), in part by Hadamard himself in 1893, and in part by von Mangoldt in 1894.

7. The discoveries of Hadamard prepared the way for rapid advançes in the theory of the distribution of primes. The prime number theorem was proved in 1896 by Hadamard himself and by de la Vallée Poussin, independently and almost simultaneously. Of the two proofs Hadamard's is the simpler, but de la Vallée Poussin (in another paper published in 1899) studied in great detail the question of closeness of approximation. His results prove conclusively (what had been foreshadowed by Chebyshev) that, for all sufficiently large values of x, $\pi(x)$ is represented more accurately by $\operatorname{li} x$ than by the function (4) (no matter what value is assigned to the constant B), and that the most favourable value of B in (4) is 1. This conflicts with Legendre's original suggestion 1·08366 for B, but this value (based on tables extending only as far as $x = 400\,000$) had long been recognised as having little more than historical interest.

The theory can now be presented in a greatly simplified form, and de la Vallée Poussin's theorems can (if desired) be proved without recourse to the theory of integral functions. This is due almost entirely to the work of Landau. The results themselves underwent no substantial change until 1921, when they were improved by Littlewood; but Littlewood's refinements lie much deeper and the proofs involve very elaborate analysis.

8. The solution just outlined may be held to be unsatisfactory in that it introduces ideas very remote from the original problem, and it is natural to ask for a proof of the prime number theorem not depending on the theory of functions of a complex variable. To this we must reply that at present no such proof is known. We can indeed go further and say that it seems unlikely that a genuinely 'real variable' proof will be discovered, at any rate so long as the theory is founded on Euler's identity. For every known proof of the prime number theorem is based on a certain property of the complex zeros of $\zeta(s)$, and this conversely is a simple consequence of the prime number theorem itself. It seems clear therefore that this property must be used (explicitly or implicitly) in any proof based on $\zeta(s)$, and it is not easy to see how this is to be done if we take account only of real values of s.

9. There is one important respect in which the theory is still very far from complete. Riemann conjectured (without any suggestion of proof) that the complex zeros of $\zeta(s)$, which (as Riemann proved) are confined to a certain infinite strip of the s-plane and lie symmetrically about the central line of this strip, are all situated on this central line. But this assertion, the now famous 'Riemann hypothesis', has never been proved or disproved, though the available evidence, both theoretical and numerical, seems to point in its favour. The truth of the Riemann hypothesis would entail considerable improvements of the theorems of de la Vallée Poussin and Littlewood on the order of magnitude of $\pi(x) - \operatorname{li} x$, but the true order cannot be decided so long as the truth of the hypothesis remains in doubt.

10. The relationship between $\pi(x)$ and $\operatorname{li} x$ is illustrated by the table on the opposite page (p. 7).[1] It will be noted at once that, for each value of x shown, $\pi(x) < \operatorname{li} x$. Until comparatively recently this inequality was believed to hold generally, and there were theoretical as well as numerical grounds for this belief; for the relation between $\pi(x)$ and $\zeta(s)$ associates $\pi(x)$ not directly with $\operatorname{li} x$, but with a more complicated expression of which the

[1] The last two entries lie outside the range of existing tables of primes, which stop a little beyond 100 000 000 [C. L. Baker and F. J. Gruenberger, *The first six million prime numbers*, The RAND Corporation, Santa Monica (published by The Microcard Foundation, Madison, Wisconsin, 1959)]. But $\pi(x)$ has been calculated for these values of x without actual enumeration of the primes; see D. H. Lehmer, *Illinois J. Math.*, 3 (1959), 381–388. The values of $\operatorname{li} x$ are given to the nearest integer. See also J. Glaisher 1, 28–38; 2, 66–103; D. N. Lehmer 1, XIII–XVI; Phragmén 2, 199–200 (footnote). [Footnote and table revised 1961.]

leading terms are $\operatorname{li} x - \frac{1}{2} \operatorname{li} x^{\frac{1}{2}}$. It was proved, however, by Littlewood (in 1914) that if we go far enough we shall eventually reach

x	$\pi(x)$	$\operatorname{li} x$	$\pi(x)/\operatorname{li} x$
1 000	168	178	0·94...
10 000	1 229	1 246	0·98...
50 000	5 133	5 167	0·993...
100 000	9 592	9 630	0·996...
500 000	41 538	41 606	0·9983...
1 000 000	78 498	78 628	0·9983...
2 000 000	148 933	149 055	0·9991...
5 000 000	348 513	348 638	0·9996...
10 000 000	664 579	664 918	0·9994...
20 000 000	1 270 607	1 270 905	0·9997...
90 000 000	5 216 954	5 217 810	0·99983...
100 000 000	5 761 455	5 762 209	0·99986...
1 000 000 000	50 847 534	50 849 235	0·99996...
10 000 000 000	455 052 512	455 055 614	0·999993...

a value of x for which $\pi(x) > \operatorname{li} x$, and that such values will recur infinitely often. Littlewood's theorem is a pure 'existence theorem', and we still know no numerical value of x for which $\pi(x) > \operatorname{li} x$. It is probable that the first such value lies far beyond the range of the above table.

There is a similar phenomenon in connection with the distribution of the odd primes between the two arithmetical progressions $4n + 1$ and $4n + 3$. If $\pi^{(1)}(x)$ and $\pi^{(3)}(x)$ denote respectively the number of primes of these two forms which do not exceed x, then $\pi^{(1)}(x)/\pi^{(3)}(x)$ tends to the limit 1 as x tends to infinity. (This theorem is of the same 'depth' as the prime number theorem, and its proof depends on the theory of functions of a complex variable.) Thus, to a first approximation, the odd primes are evenly distributed between the two progressions. But the tables show a definite preponderance of primes of the form $4n + 3$, and until 1914 all the available evidence pointed to the conclusion that (except for a short range at the beginning) $\pi^{(1)}(x) < \pi^{(3)}(x)$.[1] But Littlewood's method shows that $\pi^{(1)}(x) > \pi^{(3)}(x)$ for arbitrarily large values of x, though again it provides no numerical solution of this inequality.

[1] J. W. L. Glaisher 1. An earlier table by Scherk (*Journal für Math.*, 10 (1833), 208) is very inaccurate.

11. The present tract is devoted to a systematic study of the asymptotic relations for $\pi(x)$ discussed in outline in the foregoing sections. The theory of the Riemann zeta-function will be developed only so far as it is required for applications to $\pi(x)$. The more advanced theory of $\zeta(s)$ forms the subject of a companion volume by E. C. Titchmarsh.[1]

There are many questions relating to the distribution of primes which we are unable to discuss in this tract, either through lack of space or because the problems are as yet unsolved. To the former category belongs the general theory of the distribution of the primes among the various arithmetical progressions of given difference k, a theory associated principally with the names of Dirichlet and de la Vallée Poussin.[2] To the second category belong nearly all questions relating to the finer structure of the series of primes. The prime number theorem shows that the average interval $p_{n+1}-p_n$ between (large) consecutive primes is about $\log p_n$, but there may be wide deviations in either direction from this average. There are strong indications on the one hand that the interval reduces infinitely often to the value 2, so that there exists an infinity of pairs of primes differing only by 2 (such as 17, 19 or 10 006 427, 10 006 429); but this has not been proved. The opposite problem, that of abnormally large values of $p_{n+1}-p_n$, is also unsolved, and such indications as do exist are of a negative character. Thus, on the Riemann hypothesis, we infer easily from the results of this tract that, if ϑ is any fixed number greater than $\frac{1}{2}$, the interval is never as large as p_n^{ϑ} except possibly in a finite number of instances, and it has been conjectured that the stronger assertion with $\vartheta=\frac{1}{2}$ is also true; but the most that has actually been proved is the corresponding statement with $\vartheta>1-(33000)^{-1}$.[3]

[1] T in the bibliography at the end of this tract.

[2] For an account of this theory (and for a full treatment of the subject as a whole) we refer to the two well-known books of Landau (H and V in the bibliography).

[3] For the last result see Hoheisel 1. For an account of the delicate problems just referred to, and of related problems, see V; BC, 805–810, and the references there given; Hardy and Littlewood, 4, 5; Hardy and Littlewood, *Proc. London Math. Soc.* (2), 28 (1928), 518 (footnote); Schnirelmann 1; Landau 10.

CHAPTER I

ELEMENTARY THEOREMS

1. In this chapter we confine ourselves to theorems which can be proved without the use of the theory of functions of a complex variable. The main results are superseded by those of later chapters, but the elementary arguments are of great interest on account of their simplicity and directness.

We denote primes generally by p, and the nth prime by p_n. We denote by $\pi(x)$ the number of primes not exceeding x, where x is a positive number (not necessarily integral). We shall use the notations

$$\sum_{n \leqslant x} f(n), \quad \sum_{p > x} f(p), \quad \prod_{p} f(p), \text{ etc.}$$

(with various modifications and extensions which will be explained by the context) to indicate sums or products over all positive integers n, or all primes p, within the specified ranges; in the third example, where no range is indicated, it is understood that all primes are included. The order of the terms or factors (when relevant) is that which corresponds to increasing n or p. We adopt the general convention that an 'empty' sum (i.e. a sum containing no terms) is to have the value 0, and an 'empty' product the value 1. As examples we have (for $x > 0$)

$$[x] = \sum_{n \leqslant x} 1, \quad \pi(x) = \sum_{p \leqslant x} 1, \quad [x]! = \prod_{n \leqslant x} n,$$

where $[u]$ denotes (for any real u) the 'integral part' of u (i.e. the integer m defined by $m < u < m + 1$). We sometimes write

$$\sum_{n=1}^{x} \text{ for } \sum_{n=1}^{[x]}.$$

We use the symbols O, o, and $\sim$ (the sign of 'asymptotic equality') in the senses which are now classical. Thus

$$f(x) = O(x), \quad f(x) = o(x), \quad f(x) \sim x,$$

(as $x \to \infty$) mean respectively '$|f(x)|/x$ is less than a constant K (i.e. a number independent of x) for all sufficiently large x', '$f(x)/x \to 0$ as $x \to \infty$', and '$f(x)/x \to 1$ as $x \to \infty$'.

The notations $m \mid n$ and $m \nmid n$ mean 'm divides n' and 'm does not divide n' respectively.

2. Theorem 1. *The series and product*

$$\sum_p \frac{1}{p}, \qquad \prod_p \left(1 - \frac{1}{p}\right)^{-1}$$

are divergent.[1]

Write

$$S(x) = \sum_{p \leqslant x} \frac{1}{p}, \quad P(x) = \prod_{p \leqslant x} \left(1 - \frac{1}{p}\right)^{-1} \qquad (x > 2).$$

Since

$$1/(1-u) > (1-u^{m+1})/(1-u) = 1 + u + \ldots + u^m \qquad (0 < u < 1),$$

we have
$$P(x) > \prod_{p \leqslant x} \left(1 + \frac{1}{p} + \frac{1}{p^2} + \ldots + \frac{1}{p^m}\right),$$

where m is any positive integer. Now the product on the right, when multiplied out, is equal to $\Sigma\, 1/n$ summed over a certain set of positive integers n, and, if m is chosen so that $2^{m+1} > x$, this set will certainly include all integers from 1 to $[x]$. Hence

$$(1) \qquad P(x) > \sum_1^{[x]} \frac{1}{n} > \int_1^{[x]+1} \frac{du}{u} > \log x.$$

Since $-\log(1-u) - u < \frac{1}{2}u^2/(1-u)$ for $0 < u < 1$ (from the series for $\log(1-u)$), we have

$$\log P(x) - S(x) < \sum_{p \leqslant x} \frac{p^{-2}}{2(1-p^{-1})} < \sum_{n=2}^{\infty} \frac{1}{2n(n-1)} = \tfrac{1}{2}.$$

Hence, by (1),

$$(2) \qquad S(x) > \log\log x - \tfrac{1}{2}.$$

The inequalities (1) and (2) evidently establish the theorem.

3. Theorem 1 shows incidentally that the number of primes is infinite, or that $\pi(x)$ tends to infinity with x. We next show that $\pi(x)/x$ tends to zero.

Theorem 2. $\pi(x) = o(x)$ *as* $x \to \infty$.

Denote by $N_r(x, h)$, where $x > 0$ and h is a positive integer, the number of positive integers n not exceeding x which are divisible by h but not by any of the first r primes $p_1, \ldots, p_r$, $N_0(x, h)$ being simply the number of $n \leqslant x$ divisible by h. Then,

[1] Euler 1, Theorema 19; 2, § 279.

by considering the integers n enumerated by $N_{r-1}(x,h)$ and dividing them into two classes according as $p_r \nmid n$ or $p_r \mid n$, we see that

$$N_{r-1}(x,h) = N_r(x,h) + N_{r-1}(x, p_r h),$$

provided that $p_r \nmid h$; for in this case n is divisible both by h and by p_r if and only if it is divisible by $p_r h$. Expressing N_r in terms of N_{r-1} by means of this relation, we infer by induction that, if h is not divisible by any of $p_1, \dots, p_r$, then

$$N_r(x,h) = N_0(x,h) - \sum_i N_0(x, p_i h) + \sum_{i,j} N_0(x, p_i p_j h) - \dots,$$

where the summations are over all combinations of $p_1, \dots, p_r$, taken one, two, ..., at a time. Taking $h = 1$ and observing that $N_0(x,m) = [x/m]$, we deduce

$$N_r(x,1) = [x] - \sum_i \left[\frac{x}{p_i}\right] + \sum_{i,j} \left[\frac{x}{p_i p_j}\right] - \dots. \tag{3}$$

Now suppose $2 < \xi < x$, and let r be defined by $p_r \leqslant \xi < p_{r+1}$. Then

$$\pi(x) \leqslant r + N_r(x,1),$$

since any prime p is either one of $p_1, \dots, p_r$, or a positive integer not divisible by any of these. We substitute from (3) and omit the square brackets. This involves an error less than 1 in each term, and so a total error less than 2^r, the number of terms. Hence

$$\pi(x) < r + 2^r + x - \sum_i \frac{x}{p_i} + \sum_{i,j} \frac{x}{p_i p_j} - \dots < 2^{\xi+1} + x \prod_{p \leqslant \xi} \left(1 - \frac{1}{p}\right),$$

since $r < 2^r < 2^\xi$. Taking ξ to be a function of x such that $\xi \to \infty$ and $2^{\xi+1}/x \to 0$ as $x \to \infty$, we infer, by Theorem 1, that $\pi(x)/x \to 0$. If we take $\xi = c \log x$, where $0 < c < 1/\log 2$, and use (1), we obtain the more precise result

$$\pi(x) = O\left(\frac{x}{\log \log x}\right).$$

Euler stated that the primes are 'infinitely fewer than the integers', but his argument does not prove the assertion in the precise sense of Theorem 2.[1]

The underlying principle of the above method is that of the 'sieve of Eratosthenes'; the actual formula (3) is due to Legendre[2]. An

[1] Euler 1, Theorema 7, Corollarium 3.

[2] Legendre **1a**, 12–15; **1b**, 412–414; **2**, ii, 86–89

elaborate refinement of the method has been evolved by Brun, whereby the error arising from the omission of square brackets is greatly reduced.[1] With the aid of this refinement we can obtain the relation $\pi(x) = O(x/\log x)$, which will be proved (in a much simpler way) in § 5. The merit of Brun's method is that, while it does not profess to give a final solution of any problem, it is applicable, within its limits, to a number of problems not amenable (at present) to the analytical methods which are in general more powerful. It can be used to prove, for example, that the number of primes in *any* interval of length $x > 1$ is less than $Ax/\log x$, where A is an absolute constant.[2] Another application is to the proof of Schnirelmann's theorem that any positive integer n (>1) can be expressed as a sum of not more than k primes, where k is some absolute constant—an important contribution to the unsolved 'Goldbach's problem' concerning the possibility of expressing any even integer as a sum of two primes.[3]

4. Chebyshev's functions ϑ and ψ. It is convenient at this point to introduce Chebyshev's auxiliary functions

$$\vartheta(x) = \sum_{p \leqslant x} \log p, \quad \psi(x) = \sum_{p^m \leqslant x} \log p \qquad (x > 0),$$

where the second sum extends over every combination of a prime p with a positive integer m for which $p^m \leqslant x$.

If we group together terms of $\psi(x)$ for which m has the same value, we obtain

$$\psi(x) = \vartheta(x) + \vartheta(x^{\frac{1}{2}}) + \vartheta(x^{\frac{1}{3}}) + \ldots, \tag{4}$$

where the series on the right contains only a finite number of non-zero terms, since $\vartheta(y) = 0$ when $y < 2$. If on the other hand we group terms for which p has the same value (not exceeding x), we obtain

$$\psi(x) = \sum_{p \leqslant x} \left[\frac{\log x}{\log p}\right] \log p, \tag{5}$$

since the number of values of m associated with a given p is equal to the number of positive integers m satisfying $m \log p \leqslant \log x$, and this is $[\log x/\log p]$.

The behaviour of any one of the functions π, ϑ, ψ, for large x, can be inferred from that of any other, in virtue of the following theorem.

[1] Brun 1; Rademacher 1.

[2] Hardy and Littlewood 4, 69 (3).

[3] Schnirelmann 1; Landau 10.

Theorem 3. *The three quotients*

$$\frac{\pi(x)}{x(\log x)^{-1}}, \quad \frac{\vartheta(x)}{x}, \quad \frac{\psi(x)}{x} \tag{6}$$

have the same limits of indetermination when $x \to \infty$.

Let the upper limits (possibly $+\infty$) be Λ_1, Λ_2, Λ_3, and the lower limits λ_1, λ_2, λ_3, respectively. By (4) and (5)

$$\vartheta(x) \leqslant \psi(x) \leqslant \sum_{p \leqslant x} \frac{\log x}{\log p} \log p = \pi(x) \log x,$$

whence $\Lambda_2 \leqslant \Lambda_3 \leqslant \Lambda_1$. On the other hand, if $0 < \alpha < 1$, $x > 1$,

$$\vartheta(x) \geqslant \sum_{x^\alpha < p \leqslant x} \log p \geqslant \{\pi(x) - \pi(x^\alpha)\} \log(x^\alpha);$$

hence, since $\pi(x^\alpha) < x^\alpha$,

$$\frac{\vartheta(x)}{x} > \alpha\left(\frac{\pi(x)\log x}{x} - \frac{\log x}{x^{1-\alpha}}\right).$$

Keeping α fixed, let $x \to \infty$; since $(\log x)/x^{1-\alpha} \to 0$, we deduce that $\Lambda_2 \geqslant \alpha\Lambda_1$, whence $\Lambda_2 \geqslant \Lambda_1$, since α may be taken as near as we please to 1. This combined with the previous inequalities gives $\Lambda_2 = \Lambda_3 = \Lambda_1$. And Λ may be replaced by λ throughout the argument.

A special consequence of Theorem 3 is that, if one of the expressions (6) tends to a limit when $x \to \infty$, so do the others and the limits are all equal. Thus the three relations

$$\pi(x) \sim \frac{x}{\log x}, \quad \vartheta(x) \sim x, \quad \psi(x) \sim x,$$

the first of which is the 'prime number theorem', are equivalent.

It happens (as will appear more clearly in § 7) that, of the three functions π, ϑ, ψ, the one which arises most naturally from the analytical point of view is the one most remote from the original problem, namely ψ. For this reason it is usually most convenient to work in the first instance with ψ, and to use Theorem 3 (or more precise relations corresponding to the degree of approximation contemplated) to deduce results about π. This is a complication which seems inherent in the subject, and the reader should familiarise himself at the outset with the function ψ, which is to be regarded as the fundamental one.

We note in passing a simple arithmetical interpretation of $\psi(x)$; it is, for $x \geqslant 1$, the logarithm of the lowest common multiple of all positive integers not exceeding x.

5. The order of $\pi(x)$. We now prove that $\pi(x)$ is exactly of order $x/\log x$ when x is large.

Theorem 4. *There exist positive constants a and A such that*

$$a\frac{x}{\log x} < \pi(x) < A\frac{x}{\log x} \tag{7}$$

for all sufficiently large x.

Let Λ be the common upper limit, and λ the common lower limit, as $x\to\infty$, of the three expressions (6).

Consider the number

$$N = \frac{(n+1)(n+2)\dots(2n)}{n!},$$

where n is a positive integer. This is an integer since it is a term of the binomial expansion of $(1+1)^{2n}$; and it satisfies

$$N < 2^{2n} < (2n+1)N, \tag{8}$$

since the expansion consists of $2n+1$ positive terms of which N is the greatest. Now N is divisible by all primes p in the interval $n < p \leqslant 2n$, and therefore by their product; for these primes occur as factors in the numerator, but cannot divide any factor in the denominator. It follows that

$$N \geqslant \prod_{n<p\leqslant 2n} p.$$

This combined with the first of the inequalities (8) gives

$$2n\log 2 > \log N \geqslant \sum_{n<p\leqslant 2n} \log p = \vartheta(2n) - \vartheta(n).$$

Putting $n = 2^{r-1}$ and summing from $r=1$ to $r=m$, we deduce

$$\vartheta(2^m) < \sum_{r=1}^{m} 2^r \log 2 < 2^{m+1}\log 2.$$

Hence, if $x > 1$ and m is the integer defined by $2^{m-1} \leqslant x < 2^m$,

$$\vartheta(x) \leqslant \vartheta(2^m) < 2^{m+1}\log 2 \leqslant 4x\log 2,$$

whence $\Lambda \leqslant 4\log 2$.

To obtain an inequality in the opposite sense we use the familiar theorem that a prime p divides $m!$ exactly

$$\left[\frac{m}{p}\right] + \left[\frac{m}{p^2}\right] + \dots$$

times.[1] Since $N = (2n)!/(n!)^2$, it follows from this that

$$N = \prod_{p\leqslant 2n} p^{\nu_p},$$

[1] For the rth term $m_r = [m/p^r]$ of this sum is the number of factors in the product $1.2.\dots.m$ which are divisible by p^r, and a factor which contains p exactly r times is counted just r times in the sum $m_1+m_2+\dots$, namely once for each of the terms $m_1, m_2, \dots, m_r$.

where
$$\nu_p = \left[\frac{2n}{p}\right] + \left[\frac{2n}{p^2}\right] + \ldots - 2\left(\left[\frac{n}{p}\right] + \left[\frac{n}{p^2}\right] + \ldots\right).$$

Each of the sums on the right may be stopped at the M_pth term, where $M_p = [\log(2n)/\log p]$, since $[y] = 0$ when $0 < y < 1$. Hence
$$\nu_p = \sum_{r=1}^{M_p}\left(\left[\frac{2n}{p^r}\right] - 2\left[\frac{n}{p^r}\right]\right) \leqslant M_p,$$

since $[2y] - 2[y]$ is always 0 or 1. But, by (5),
$$e^{\psi(2n)} = \prod_{p \leqslant 2n} p^{M_p}.$$

Thus $N \mid e^{\psi(2n)}$, whence, by the second of the inequalities (8),
$$2n \log 2 - \log(2n+1) < \log N \leqslant \psi(2n).$$

Hence, if $x > 2$ and $n = [\tfrac{1}{2}x]$, we have
$$\psi(x) \geqslant \psi(2n) > (x-2)\log 2 - \log(x+1),$$

whence $\lambda \geqslant \log 2$.

The inequalities $\log 2 \leqslant \lambda \leqslant \Lambda \leqslant 4 \log 2$ evidently imply the result stated.

By a slight modification of the argument it is easy to show that $\log 2 \leqslant \lambda \leqslant \Lambda \leqslant 2 \log 2$. But Chebyshev[1], to whom Theorem 4 is due, had a definite application in view—the proof of 'Bertrand's postulate' that, if $n > 6$, there always exists a prime p satisfying $\tfrac{1}{2}n < p \leqslant n - 2$—and for this he required the inequality $\Lambda < 2\lambda$. This he secured by proving that
$$h \leqslant \lambda \leqslant \Lambda \leqslant H,$$

where
$$h = \log \frac{2^{\frac{1}{2}} 3^{\frac{1}{3}} 5^{\frac{1}{5}}}{30^{\frac{1}{30}}} = 0{\cdot}921\ldots, \quad H = \tfrac{6}{5}h = 1{\cdot}105\ldots.$$

Still narrower limits were obtained by subsequent writers[2], but all attempts to prove the prime number theorem ($\lambda = \Lambda = 1$) by these methods were unsuccessful.[3]

Theorem 2 is, of course, included in Theorem 4, and Theorem 1 also is an easy deduction. For the inequality $\pi(x) > ax/\log x$ shows, first that the number of primes is infinite, and then that, for sufficiently large n,
$$n = \pi(p_n) > ap_n/\log p_n > p_n^{\frac{1}{2}};$$

thus
$$ap_n < n \log p_n < 2n \log n \quad (n > n_0),$$

and Σp_n^{-1} diverges by comparison with $\Sigma(n \log n)^{-1}$.

[1] Chebyshev 3.

[2] Sylvester 1, 2, 3; H, i, 87–95. See also Schur 1, 2; Breusch 1.

[3] Cf. H, ii, 597–604.

6. Euler's identity. We now introduce the fundamental identity of the theory of primes. This is a particular case of the following theorem.[1]

Theorem 5. *Let $f(n)$ be a 'multiplicative' function of the positive integral variable n, i.e. one which is not identically* 0 *and has the property that*

$$f(m)f(n) = f(mn) \tag{9}$$

whenever m and n are prime to one another. Then

$$\sum_1^\infty f(n) = \prod_p \{1 + f(p) + f(p^2) + \ldots\}, \tag{10}$$

provided that the series on the left is absolutely convergent, in which case the product is also absolutely convergent.

If $f(n)$ is 'completely multiplicative', i.e. if (9) *holds for all m and n, co-prime or not, we can write*

$$\sum_1^\infty f(n) = \prod_p \frac{1}{1-f(p)}. \tag{11}$$

We observe first that $f(1) = 1$; for $f(n)f(1) = f(n)$ by (9), and n may be chosen so that $f(n) \neq 0$. Now consider the product

$$P(x) = \prod_{p \leq x} \{1 + f(p) + f(p^2) + \ldots\} \qquad (x > 2).$$

The number of factors is finite, and each is an absolutely convergent series since $\Sigma |f(n)|$ is convergent. Hence, by Cauchy's theorem on multiplication of series, we may multiply out, and arrange the terms of the formal product in any order as a simple series, which will be absolutely convergent. By the multiplicative property of $f(n)$ and the fundamental theorem of arithmetic that any $n > 1$ can be expressed uniquely as a product of prime factors, we obtain

$$P(x) = \Sigma f(n'),$$

where the summation is over all positive integers n' having no prime factor greater than x. Hence, if S is the sum of the series on the left of (10),

$$P(x) - S = -\Sigma f(n''),$$

[1] Used (formally) in a number of special cases by Euler (1, Theorema 8; 2, Caput 15).

where n'' runs through all positive integers having at least one prime factor greater than x. Since every n'' must be greater than x, it follows that

$$|P(x)-S|<\Sigma|f(n'')|<\sum_{n>x}|f(n)|.$$

When $x\to\infty$ the expression on the right tends to 0, since $\Sigma|f(n)|$ is convergent; hence $P(x)\to S$. This proves (10); the product is absolutely convergent since

$$\sum_{p\leqslant x}|f(p)+f(p^2)+\dots|<\sum_{p\leqslant x}\{|f(p)|+|f(p^2)|+\dots\}<\sum_{n=2}^{\infty}|f(n)|.$$

If (9) holds generally, then

$$1+f(p)+f(p^2)+\dots=1+f(p)+(f(p))^2+\dots=\frac{1}{1-f(p)},$$

the series being already known to be convergent; this gives (11).

7. Fundamental formulae. In Theorem 5 take $f(n)=n^{-s}$, where $s>1$. The conditions for (11) are satisfied and we obtain

$$\sum_1^{\infty}n^{-s}=\prod_p\frac{1}{1-p^{-s}}\qquad(s>1). \tag{12}$$

Supposing always that $s>1$, we write

$$\zeta(s)=\sum_1^{\infty}n^{-s}.$$

Taking logarithms in (12) and expanding, we obtain

$$\log\zeta(s)=-\sum_p\log(1-p^{-s})=\sum_{p,\,m}\frac{1}{mp^{ms}}, \tag{13}$$

where p runs through all primes and m through all positive integers. By differentiation,

$$-\frac{\zeta'(s)}{\zeta(s)}=\sum_p\frac{p^{-s}\log p}{1-p^{-s}}=\sum_{p,\,m}\frac{\log p}{p^{ms}}.$$

This may be written as

$$-\frac{\zeta'(s)}{\zeta(s)}=\sum_1^{\infty}\frac{\Lambda(n)}{n^s}, \tag{14}$$

where $\Lambda(n)$ is $\log p$ if n is a (positive) power of a prime p, and is 0 otherwise. These transformations are legitimate when $s>1$, since the double series have their terms positive, and the simple series are uniformly convergent for $s\geqslant 1+\delta$, where δ is any fixed positive number.

If $\psi(x)$ is Chebyshev's function defined in § 4, we have

$$(15)\qquad \psi(x) = \sum_{p^m \leqslant x} \log p = \sum_{n \leqslant x} \Lambda(n).$$

Thus $\psi(x)$ arises naturally out of the above formulae as the sum of the first $[x]$ coefficients c_n in the expansion of $-\zeta'(s)/\zeta(s)$ as a 'Dirichlet's series' $\Sigma c_n n^{-s}$; or $\psi(x)$ is the 'sum-function' of the coefficients of the 'generating function' (14). The function similarly associated with $\log \zeta(s)$ is

$$(16)\qquad \Pi(x) = \sum_{p^m \leqslant x} \frac{1}{m} = \pi(x) + \tfrac{1}{2}\pi(x^{\frac{1}{2}}) + \tfrac{1}{3}\pi(x^{\frac{1}{3}}) + \dots.$$

From (14) and (15) we can deduce

$$(17)\qquad -\frac{\zeta'(s)}{\zeta(s)} = s\int_1^\infty \frac{\psi(x)}{x^{s+1}}\,dx \qquad (s > 1);$$

and from (13) and (16) that

$$(18)\qquad \log \zeta(s) = s\int_1^\infty \frac{\Pi(x)}{x^{s+1}}\,dx \qquad (s > 1).$$

The proofs depend on the following theorem, a convenient form of the formula for 'partial summation' (Abel's identity) which will find numerous applications in the sequel.

Theorem A.[1] *Let $\lambda_1, \lambda_2, \dots$ be a real sequence which increases (in the wide sense) and has the limit infinity, and let*

$$C(x) = \sum_{\lambda_n \leqslant x} c_n,$$

where the c_n may be real or complex, and the notation indicates a summation over the (finite) set of positive integers n for which $\lambda_n \leqslant x$. Then, if $X \geqslant \lambda_1$ and $\phi(x)$ has a continuous derivative, we have

$$(19)\qquad \sum_{\lambda_n \leqslant X} c_n \phi(\lambda_n) = -\int_{\lambda_1}^X C(x)\phi'(x)\,dx + C(X)\phi(X).$$

If, further, $C(X)\phi(X) \to 0$ as $X \to \infty$, then

$$\sum_1^\infty c_n \phi(\lambda_n) = -\int_{\lambda_1}^\infty C(x)\phi'(x)\,dx,$$

provided that either side is convergent.

[1] Theorems which do not refer explicitly to primes, or to the special functions associated with them, will be distinguished by letters instead of numbers.

Denoting by S the left-hand side of (19), we have

$$C(X)\phi(X) - S = \sum_{\lambda_n \leqslant X} c_n\{\phi(X) - \phi(\lambda_n)\} = \sum_{\lambda_n \leqslant X} \int_{\lambda_n}^{X} c_n \phi'(x)\, dx$$

$$= \int_{\lambda_1}^{X} \sum_{\lambda_n \leqslant x} c_n \phi'(x)\, dx = \int_{\lambda_1}^{X} C(x)\phi'(x)\, dx,$$

the desired result. The interchange of the order of summation and integration presents no theoretical difficulty since only finite ranges are involved; and the result is formally correct, since the combined range of summation and integration is defined by $\lambda_1 \leqslant \lambda_n \leqslant x \leqslant X$.[1] This proves (19), and the second formula follows on making $X \to \infty$.

Note. We have assumed for simplicity that the suffixes in λ_n and c_n run from 1 onwards. But it is sometimes convenient to adopt a different starting-point (e.g. 0 or 2); it is clear that the formulae, suitably modified, are still valid.

Theorem A may also be deduced directly from Abel's identity. Another proof is by partial integration in a Stieltjes integral; this is perhaps the most fundamental point of view.

To deduce (17) (for example) from Theorem A, take

$$\lambda_n = n, \quad c_n = \Lambda(n), \quad \phi(x) = x^{-s},$$

and observe that, for fixed $s > 1$,

$$C(X)\phi(X) = \psi(X)X^{-s} = O(X^{1-s}\log X) = o(1)$$

as $X \to \infty$.

The above formulae are fundamental in all that follows. The notations $\zeta(s)$, $\Lambda(n)$, $\Pi(x)$, introduced in this section, will be adhered to throughout.

Theorem 1 evidently rests on the same principle as Euler's identity, and we may base the proof directly on this identity by making $s \to 1 + 0$ in (12). Or we may simply take $f(n) = 1/n$ in (10); for it is easily shown that, when $f(n) \geqslant 0$, the convergence of the product implies that of the series. A less obvious remark is that the formulae of § 5 are also connected

[1] We use, in fact, the formula

$$\sum_{\lambda_n \leqslant X} \int_{\lambda_1}^{X} f_n(x)\, dx = \int_{\lambda_1}^{X} \sum_{\lambda_n \leqslant X} f_n(x)\, dx,$$

where $f_n(x)$ is equal to 0 for $x < \lambda_n$ and to $c_n \phi'(x)$ for $x \geqslant \lambda_n$.

with Euler's identity. In fact, the theorem on divisibility of factorials which was used in the second part of the proof of Theorem 4 (and might also have been used in the first part) is equivalent to

$$(20) \qquad m! = \prod_{p \leqslant m} p^{\left[\frac{m}{p}\right] + \left[\frac{m}{p^2}\right] + \dots},$$

or (on taking logarithms) to

$$\sum_{n=1}^{m} \log n = \sum_{p^r \leqslant m} \left[\frac{m}{p^r}\right] \log p = \sum_{n_1 \leqslant m} \left[\frac{m}{n_1}\right] \Lambda(n_1) = \sum_{n_1 n_2 \leqslant m} \Lambda(n_1),$$

and therefore asserts the equality of the sum of the first m coefficients on the left and right of the identity

$$\sum_{n=1}^{\infty} \frac{\log n}{n^s} = -\zeta'(s) = -\frac{\zeta'(s)}{\zeta(s)} \cdot \zeta(s) = \sum_{n_1=1}^{\infty} \frac{\Lambda(n_1)}{n_1^s} \cdot \sum_{n_2=1}^{\infty} \frac{1}{n_2^s},$$

when the product on the right is expressed as a Dirichlet's series $\Sigma c_n n^{-s}$. The formal multiplication is legitimate when $s > 1$, and the process of 'equating coefficients' is permissible for Dirichlet's series as for power series.[1] Thus (20) may be deduced from (14) and so from Euler's identity.

8. We can now prove the theorem that, *if* $\pi(x)/(x/\log x)$ *tends to a limit when* $x \to \infty$, *the limit must be* 1. This will evidently follow from

Theorem 6. *We have*

$$\underline{\lim}\, \pi(x) \Big/ \frac{x}{\log x} \leqslant 1 \leqslant \overline{\lim}\, \pi(x) \Big/ \frac{x}{\log x}$$

when $x \to \infty$.[2]

Write $f(s) = -\zeta'(s)/\zeta(s)$, and let

$$\overline{\underline{\lim}}_{x \to \infty} \frac{\psi(x)}{x} = \begin{cases} \Lambda \\ \lambda \end{cases}, \quad \overline{\underline{\lim}}_{s \to 1+0} (s-1) f(s) = \begin{cases} \Lambda' \\ \lambda' \end{cases}.$$

If Λ is not $+\infty$[3], choose $B > \Lambda$. Then we have $\psi(x)/x < B$ for all $x > x_0 = x_0(B)$, and we may choose $x_0 > 1$. We deduce, by (17), that, for $s > 1$,

$$f(s) = s \int_1^{\infty} \frac{\psi(x)}{x^{s+1}} dx < s \int_1^{x_0} \frac{\psi(x)}{x^{s+1}} dx + s \int_{x_0}^{\infty} \frac{B}{x^s} dx$$

$$= s \int_1^{x_0} \frac{\psi(x) - Bx}{x^{s+1}} dx + \frac{sB}{s-1} < s \int_1^{x_0} \frac{|\psi(x) - Bx|}{x^2} dx + \frac{sB}{s-1},$$

so that $\qquad (s-1) f(s) < s(s-1) K + sB,$

[1] HR, Theorems 53 and 6.

[2] Chebyshev 1, 2, Theorem II ($n = 1$) (cf. H, i, 16).

[3] We know of course from Theorem 4 that Λ is finite, but there is no need to use this fact.

where $K = K(B, x_0) = K(B)$[1]. Making $s \to 1+0$ we deduce that $\Lambda' < B$, and this, being true for every $B > \Lambda$, implies that $\Lambda' < \Lambda$. And this inequality is trivial (with an obvious interpretation) if Λ is $+\infty$. We can prove similarly that $\lambda' > \lambda$. Hence

$$(21) \qquad \lambda \leqslant \lambda' \leqslant \Lambda' \leqslant \Lambda.$$

Now since x^{-s} is a decreasing function of x (for fixed $s > 1$) we have

$$\int_1^\infty \frac{dx}{x^s} < \sum_1^\infty \frac{1}{n^s} < 1 + \int_1^\infty \frac{dx}{x^s},$$

i.e.

$$\frac{1}{s-1} < \zeta(s) < \frac{s}{s-1};$$

hence $(s-1)\zeta(s) \to 1$ as $s \to 1+0$. And since $(\log x)/x^s$ is a decreasing function of x for $x > e$ (if $s > 1$), we see in a similar way that, when $s \to 1+0$,

$$-\zeta'(s) = \sum_1^\infty \frac{\log n}{n^s} = \int_1^\infty \frac{\log x}{x^s} dx + O(1)$$

$$= \frac{1}{(s-1)^2} \int_0^\infty y e^{-y} dy + O(1) = \frac{1}{(s-1)^2} + O(1)$$

by the substitution $x^{s-1} = e^y$. Hence

$$(s-1) f(s) = \frac{-(s-1)^2 \zeta'(s)}{(s-1)\zeta(s)} \to 1$$

as $s \to 1+0$, so that $\lambda' = \Lambda' = 1$. This combined with (21) gives $\lambda \leqslant 1 \leqslant \Lambda$, which, by Theorem 3, is equivalent to the result stated.

Chebyshev went a good deal further in the same direction. His results indicate that the most favourable value of B in Legendre's formula $x/(\log x - B)$ is 1, and that the function $\operatorname{li} x$ (which he introduced independently of Gauss) is preferable, as an approximation to $\pi(x)$, to any rational function of x and $\log x$. Stated precisely, his theorem on the value of B, for example, is that, if the function $B(x)$ defined by

$$\pi(x) = \frac{x}{\log x - B(x)}$$

tends to a limit when $x \to \infty$, the limit must be 1.[2] These results have now been completed by the proof of the existence of the relevant limits. They are easy deductions from the results of Chapter III, § 12, pp. 65–66.

[1] That is to say, K is a number depending in the first instance on B and x_0, and so ultimately only on B, since x_0 depends only on B; K is a 'constant' in the sense that it is independent of the main variables x and s.

[2] Chebyshev 1, 2, Theorem III; H, i, 140–150.

9. We conclude this chapter with a number of interesting asymptotic formulae, due to Mertens, for sums and products involving primes.[1] The proofs depend on Chebyshev's theorem $\psi(x) = O(x)$ (which is equivalent to the second inequality of Theorem 4), and on the fact, established in the course of the proof of Theorem 6, that $(s-1)\zeta(s) \to 1$ as $s \to 1+0$.

Theorem 7. *When* $x \to \infty$

$$\sum_{p \leqslant x} \frac{\log p}{p} = \log x + O(1), \tag{22}$$

$$\sum_{p \leqslant x} \frac{1}{p} = \log\log x + B + O\left(\frac{1}{\log x}\right), \tag{23}$$

$$\prod_{p \leqslant x} \left(1 - \frac{1}{p}\right) \sim \frac{e^{-C}}{\log x}, \tag{24}$$

where B and C are constants, of which C is Euler's constant.

By (20)

$$\log m! = \sum_{p^r \leqslant m} \left[\frac{m}{p^r}\right] \log p = \sum_{n \leqslant m} \left[\frac{m}{n}\right] \Lambda(n). \tag{25}$$

By Stirling's theorem, $\log m! = m \log m + O(m)$ as $m \to \infty$. The omission of the square brackets on the right of (25) involves an error at most $\psi(m) = O(m)$. Using these approximations and dividing through by m, we obtain

$$\sum_{n \leqslant m} \frac{\Lambda(n)}{n} = \log m + O(1),$$

and m may now be replaced by a continuous variable x. This result is equivalent to (22), since

$$\left| \sum_{n \leqslant x} \frac{\Lambda(n)}{n} - \sum_{p \leqslant x} \frac{\log p}{p} \right| < \sum_{p \leqslant x} \left(\frac{1}{p^2} + \frac{1}{p^3} + \dots\right) \log p < \sum_p \frac{\log p}{p(p-1)},$$

the last series being convergent.

Next, writing

$$A(x) = \sum_{p_n \leqslant x} \frac{\log p_n}{p_n} = \sum_{p_n \leqslant x} a_n, \qquad B(x) = \sum_{p_n \leqslant x} \frac{1}{p_n} = \sum_{p_n \leqslant x} b_n,$$

we have, for $x > 2$,

$$B(x) = \sum_{p_n \leqslant x} \frac{a_n}{\log p_n} = \int_2^x \frac{A(u)}{u(\log u)^2} du + \frac{A(x)}{\log x},$$

by Theorem A. By (22), $A(x) = \log x + r(x)$, where $|r(x)|$ is less than a constant K for all $x > 2$. Hence

$$B(x) = \int_2^x \frac{du}{u \log u} + \int_2^x \frac{r(u)\, du}{u(\log u)^2} + 1 + \frac{r(x)}{\log x}$$

$$= \log\log x + B + R(x), \tag{26}$$

[1] Mertens 1; see also Hardy 2.

where B is a certain constant, and

$$|R(x)| = \left| -\int_x^\infty \frac{r(u)\,du}{u(\log u)^2} + \frac{r(x)}{\log x} \right| < \int_x^\infty \frac{K\,du}{u(\log u)^2} + \frac{K}{\log x} = \frac{2K}{\log x},$$

for $x \geqslant 2$. This proves (23).

We shall now determine B by investigating in two ways the behaviour of $g(s) = \Sigma p^{-s}$ when $s \to 1+0$. On the one hand we have, for $s > 1$,

$$g(s) = \sum_1^\infty b_n p_n^{1-s} = (s-1)\int_2^\infty B(x)\, x^{-s}\,dx,$$

by Theorem A, since $B(X)X^{1-s} \to 0$ as $X \to \infty$. Substituting from (26) we obtain

$$g(s) = (s-1)\int_2^\infty \frac{\log\log x}{x^s}\,dx + (s-1)\int_2^\infty \frac{B}{x^s}\,dx + (s-1)\int_2^\infty \frac{R(x)}{x^s}\,dx$$
$$= I_1 + I_2 + I_3,$$

say. Since

$$R(x) = O(1/\log x) = o(1),$$

we have $|R(x)| < \epsilon$ for all $x > x_0 = x_0(\epsilon)\,(>2)$. Hence

$$|I_3| < (s-1)\int_2^{x_0} |R(x)|\,dx + \epsilon < 2\epsilon$$

for $1 < s < s_0 = s_0(\epsilon, x_0) = s_0(\epsilon)$; thus $I_3 \to 0$ as $s \to 1+0$. In I_1 and I_2 we replace the lower limit of integration by 1; this involves an error $o(1)$. We then make the substitution $x^{s-1} = e^y$ in I_1. We thus obtain

$$g(s) = \int_0^\infty e^{-y} \log \frac{y}{s-1}\,dy + B + o(1)$$

$$= -C - \log(s-1) + B + o(1) \tag{27}$$

when $s \to 1+0$, where C is Euler's constant[1]. On the other hand, by (13),

$$\log(s-1) + g(s) = \log\{(s-1)\zeta(s)\} + \sum_p \left\{\log\left(1 - \frac{1}{p^s}\right) + \frac{1}{p^s}\right\}$$

$$\to \sum_p \left\{\log\left(1 - \frac{1}{p}\right) + \frac{1}{p}\right\} \tag{28}$$

as $s \to 1+0$, since $(s-1)\zeta(s) \to 1$ and the series on the right is uniformly convergent for $s \geqslant 1$ (by comparison with Σp^{-2}). Comparing (27) and (28) we deduce

$$B = C + \sum_p \left\{\log\left(1 - \frac{1}{p}\right) + \frac{1}{p}\right\}.$$

Substituting for B into (23) and transposing, we obtain

$$\sum_{p \leqslant x} \frac{1}{p} - \sum_p \left\{\log\left(1 - \frac{1}{p}\right) + \frac{1}{p}\right\} = \log\log x + C + O\left(\frac{1}{\log x}\right).$$

[1] The formula

$$\int_0^\infty e^{-y} \log y\,dy = -C$$

may be obtained from Euler's integral for $\Gamma(z)$ by differentiating (under the integral sign) and putting $z = 1$.

In the second sum on the left we may replace the infinite range of summation by the finite range $p < x$, with an error

$$O\left(\sum_{p>x}\frac{1}{p^2}\right)=O\left(\sum_{n>x}\frac{1}{n^2}\right)=O\left(\frac{1}{x}\right).$$

We thus obtain

$$-\sum_{p\leqslant x}\log\left(1-\frac{1}{p}\right)=\log\log x+C+O\left(\frac{1}{\log x}\right),$$

or

$$\prod_{p\leqslant x}\left(1-\frac{1}{p}\right)=\frac{e^{-C}}{\log x}e^{O\left(\frac{1}{\log x}\right)}=\frac{e^{-C}}{\log x}\left\{1+O\left(\frac{1}{\log x}\right)\right\},$$

which implies (24).

The relation (24) is important as showing that various 'probability' arguments which have been put forward tentatively from time to time lead to results which are certainly false.[1]

[1] Hardy and Littlewood 4, 32–37.

CHAPTER II

THE PRIME NUMBER THEOREM

1. The Riemann zeta-function. In this chapter our aim is to prove the prime number theorem, and for this we must study Riemann's function

$$\zeta(s)=\sum_{1}^{\infty}\frac{1}{n^s} \tag{1}$$

as a function of the complex variable s. We shall confine ourselves here to such properties as are required for the immediate application, reserving for the next chapter the systematic theory of the zeta-function.

We write
$$s=\sigma+ti,$$
and define x^s generally, for $x>0$, as $\exp(s\log x)$, where $\log x$ has its real determination. Then $|n^s|=n^\sigma$, and the series (1) is uniformly convergent (by Weierstrass's test) for $\sigma>1+\delta$, where δ is any fixed positive number. Hence $\zeta(s)$ is regular in the half-plane $\sigma>1$, and its derivatives may be calculated by term-by-term differentiation.[1]

The formulae of Chapter I, § 7 (pp. 17–20) hold throughout the half-plane $\sigma>1$. For the formal transformations evidently retain their validity, and for the theoretical justification it is sufficient to observe that the various series are majorised by the same series with σ in place of s. The branch of $\log\zeta(s)$ to be taken in (13) and (18) is the one which is real on the real axis. The

[1] By Weierstrass's classical theorem on uniformly convergent sequences, $f_n(s)$, of regular functions. We shall have occasion to use also the analogous theorem for $f_X(s)$, where X is a continuous variable. In the next section, for example, we are concerned with

$$f(s)=\int_1^\infty\frac{(x)}{x^{s+1}}\,dx=\lim_{X\to\infty}f_X(s),\qquad f_X(s)=\int_1^X\frac{(x)}{x^{s+1}}\,dx,$$

where $(x)=x-[x]$. In this and similar cases the regularity of $f_X(s)$ (for fixed X) in the whole s-plane may be established by expanding the integrand in powers of s and integrating term-by-term, or may be inferred from general theorems. Either argument shows that the derivatives of $f_X(s)$ may be obtained by differentiation under the integral sign; the same is therefore true, by Weierstrass's theorem, of $f(s)$ in a domain of uniform convergence. We often (as here) have to apply the theorems to a smaller domain than the one in which we are ultimately interested.

function $\zeta(s)$ does not vanish in $\sigma > 1$, since Euler's product $\zeta(s) = \Pi(1-p^{-s})^{-1}$ is convergent and no factor vanishes; thus $\log \zeta(s)$ has no singularities in $\sigma > 1$.

2. Analytic continuation. So far $\zeta(s)$ is defined only in the half-plane $\sigma > 1$, and we must consider the question of its analytic continuation. For the present we content ourselves with the following special result.

Theorem 8. *The function $\zeta(s)$, defined for $\sigma > 1$ by (1), admits of analytic continuation over the half-plane $\sigma > 0$, as a single-valued function having as its only singularity in this half-plane a simple pole with residue 1 at $s = 1$.*

By Theorem A (p. 18), with $\lambda_n = n$, $c_n = 1$, $\phi(x) = x^{-s}$,

$$\sum_{n \leqslant X} \frac{1}{n^s} = s \int_1^X \frac{[x]}{x^{s+1}} dx + \frac{[X]}{X^s},$$

if $X \geqslant 1$. Writing $[x] = x - (x)$, so that $0 \leqslant (x) < 1$, we obtain

$$(2) \qquad \sum_{n \leqslant X} \frac{1}{n^s} = \frac{s}{s-1} - \frac{s}{(s-1)X^{s-1}} - s\int_1^X \frac{(x)}{x^{s+1}} dx + \frac{1}{X^{s-1}} - \frac{(X)}{X^s}.$$

Since $|1/X^{s-1}| = 1/X^{\sigma-1}$ and $|(X)/X^s| < 1/X^\sigma$, we deduce, making $X \to \infty$,

$$(3) \qquad \zeta(s) = \frac{s}{s-1} - s\int_1^\infty \frac{(x)}{x^{s+1}} dx,$$

if $\sigma > 1$. Since $|(x)/x^{s+1}| < 1/x^{\sigma+1}$, the last integral is uniformly convergent for $\sigma > \delta$, where δ is any fixed positive number, and therefore represents a regular function of s in $\sigma > 0$ (p. 25, footnote). This proves the theorem, the equation (3) providing the continuation of $\zeta(s)$ over the half-plane $\sigma > 0$.

3. In any fixed half-plane $\sigma \geqslant 1 + \delta > 1$, $\zeta(s)$ is bounded, since $|\zeta(s)| \leqslant \zeta(\sigma) \leqslant \zeta(1+\delta)$. The formulae of § 2 can be made to give important information about the order of magnitude of $|\zeta(s)|$ in the neighbourhood of the line $\sigma = 1$ and to the left of it. We are primarily interested in points $s = \sigma + ti$ with large $|t|$, and we may confine ourselves to the upper half-plane $t > 0$, since $\zeta(\sigma + ti)$ and $\zeta(\sigma - ti)$ are conjugate complex numbers, as appears from equation (3).

Here and in all that follows we shall use the letter A (and sometimes a), with or without distinguishing mark, to denote a positive absolute constant whose value we leave unspecified. We shall denote by $A(\alpha, \beta, \ldots)$ a positive number depending on the parameters $\alpha, \beta, \ldots$ shown explicitly and on no others. Finally we use the recognised extensions of the O-notation to functions involving parameters; thus we say that $f(\sigma, t) = O(t^{\sigma})$, uniformly in a stated range of σ, as $t \to \infty$, if $|f(\sigma, t)| < Kt^{\sigma}$ for all $t > t_0$ and all σ in the range, where K and t_0 are independent of both t and σ.

Theorem 9. *We have*

$$|\zeta(s)| < A \log t \qquad (\sigma \geqslant 1,\ t \geqslant 2), \tag{4}$$

$$|\zeta'(s)| < A \log^2 t \qquad (\sigma \geqslant 1,\ t \geqslant 2), \tag{5}$$

$$|\zeta(s)| < A(\delta)\, t^{1-\delta} \qquad (\sigma \geqslant \delta,\ t \geqslant 1), \tag{6}$$

if $0 < \delta < 1$.

By (2) and (3) we have, for $\sigma > 0$, $t \geqslant 1$, $X \geqslant 1$,

$$\zeta(s) - \sum_{n \leqslant X} \frac{1}{n^s} = \frac{1}{(s-1)X^{s-1}} + \frac{(X)}{X^s} - s\int_X^{\infty} \frac{(x)\,dx}{x^{s+1}}. \tag{7}$$

Hence

$$|\zeta(s)| < \sum_{n \leqslant X} \frac{1}{n^{\sigma}} + \frac{1}{tX^{\sigma-1}} + \frac{1}{X^{\sigma}} + |s| \int_X^{\infty} \frac{dx}{x^{\sigma+1}}$$

$$< \sum_{n \leqslant X} \frac{1}{n^{\sigma}} + \frac{1}{tX^{\sigma-1}} + \frac{1}{X^{\sigma}} + \left(1 + \frac{t}{\sigma}\right)\frac{1}{X^{\sigma}},$$

since $|s| < \sigma + t$. If $\sigma \geqslant 1$,

$$|\zeta(s)| < \sum_{n \leqslant X} \frac{1}{n} + \frac{1}{t} + \frac{1}{X} + \frac{1+t}{X} < (\log X + 1) + 3 + \frac{t}{X},$$

since $t \geqslant 1$, $X \geqslant 1$. Taking $X = t$ we obtain (4).

If $\sigma \geqslant \eta$, where $0 < \eta < 1$,

$$|\zeta(s)| < \sum_{n \leqslant X} \frac{1}{n^{\eta}} + \frac{1}{tX^{\eta-1}} + \left(2 + \frac{t}{\eta}\right)\frac{1}{X^{\eta}}$$

$$< \int_0^{[X]} \frac{dx}{x^{\eta}} + \frac{X^{1-\eta}}{t} + \frac{3t}{\eta X^{\eta}} < \frac{X^{1-\eta}}{1-\eta} + X^{1-\eta} + \frac{3t}{\eta X^{\eta}}.$$

Taking $X = t$, as before, we deduce

$$|\zeta(s)| < t^{1-\eta}\left(\frac{1}{1-\eta} + 1 + \frac{3}{\eta}\right) \qquad (\sigma \geqslant \eta,\ t \geqslant 1), \tag{8}$$

which implies (6).

The inequality (5) may be deduced in a similar way from the formula resulting from the differentiation of (7). Or we may argue as follows. Let $s_0 = \sigma_0 + t_0 i$ be any point in the region $\sigma \geqslant 1$, $t \geqslant 2$, and C a circle with centre s_0 and radius $\rho < \frac{1}{2}$. Then

$$|\zeta'(s_0)| = \left| \frac{1}{2\pi i} \int_C \frac{\zeta(s)\, ds}{(s-s_0)^2} \right| \leqslant \frac{M}{\rho},$$

where M is the maximum of $|\zeta(s)|$ on C. Now $\sigma \geqslant \sigma_0 - \rho \geqslant 1 - \rho$ and $1 < t < 2t_0$ at all points s on C. Hence by (8)

$$M < (2t_0)^\rho \left(\frac{1}{\rho} + 1 + \frac{3}{1-\rho} \right) < \frac{10 t_0{}^\rho}{\rho},$$

since $\rho < 1 - \rho < 1$, $2^\rho < 2$. Hence

$$|\zeta'(s_0)| < 10 t_0{}^\rho / \rho^2.$$

Take $\rho = 1/(\log t_0 + 2)$. Then $t_0{}^\rho = e^{\rho \log t_0} < e$, so that

$$|\zeta'(s_0)| < 10e(\log t_0 + 2)^2.$$

This implies (5) since s_0 is any point in $\sigma \geqslant 1$, $t \geqslant 2$.

The inequalities (4), (5), (6) make no claim to be the best possible of their kind, and a similar remark applies to other inequalities involving $\zeta(s)$ which will be obtained in the course of this tract. Further refinements of the inequalities for $\zeta(s)$ are of little importance in the theory of primes, since they are lost in the passage from $\zeta(s)$ to the logarithm or logarithmic derivative, the functions which bear directly on the theory of primes. But the problem of the true order of $\zeta(\sigma + ti)$ as a function of t is one of the most important in the pure theory of $\zeta(s)$. It is also one of the most difficult, and its solution is very far from complete.[1]

4. Zeros. In the application of $\zeta(s)$ to the theory of primes it is natural to expect that zeros of $\zeta(s)$ will play an important part, since these are singularities of the fundamental functions $\log \zeta(s)$ and $-\zeta'(s)/\zeta(s)$. Euler's product $\zeta(s) = \Pi(1 - p^{-s})^{-1}$ shows (as we remarked in § 1) that there are no zeros in the half-plane $\sigma > 1$, but it gives no direct information about points not belonging to this domain. We must first establish the following theorem.

Theorem 10. *$\zeta(s)$ has no zeros on the line $\sigma = 1$. Further*

(9) $$\frac{1}{\zeta(s)} = O\{(\log t)^A\},$$

uniformly for $\sigma \geqslant 1$, as $t \to \infty$.

[1] See T, Chapters I, V, VI.

We base the proof on the elementary inequality

$$(10) \qquad 3+4\cos\theta+\cos 2\theta \geqslant 0,$$

which holds for all real θ, since the left-hand side is $2(1+\cos\theta)^2$.

By equation (13), p. 17, we have, for $\sigma > 1$,

$$\log|\zeta(\sigma+ti)| = \Re \sum_{n=2}^{\infty} c_n n^{-\sigma-ti} = \sum_{n=2}^{\infty} c_n n^{-\sigma}\cos(t\log n),$$

where c_n is $1/m$ if n is the mth power of a prime, and 0 otherwise. Hence[1]

$$\log|\zeta^3(\sigma)\zeta^4(\sigma+ti)\zeta(\sigma+2ti)|$$
$$= \Sigma c_n n^{-\sigma}\{3+4\cos(t\log n)+\cos(2t\log n)\} \geqslant 0,$$

by (10), since $c_n \geqslant 0$. Thus

$$(11) \quad \{(\sigma-1)\zeta(\sigma)\}^3 \left|\frac{\zeta(\sigma+ti)}{\sigma-1}\right|^4 |\zeta(\sigma+2ti)| \geqslant \frac{1}{\sigma-1} \quad (\sigma>1).$$

This shows that the point $1+ti$ ($t \gtrless 0$) cannot be a zero of $\zeta(s)$. For, if it were, then, since $\zeta(s)$ is regular at the points $1+ti$ and $1+2ti$, and has a simple pole (with residue 1) at the point 1, the left-hand side would tend to a finite limit (viz. $|\zeta'(1+ti)|^4|\zeta(1+2ti)|$), and the right-hand side to infinity, when $\sigma \to 1+0$. This proves the first part of the theorem.

In proving the second part we may suppose $1 \leqslant \sigma \leqslant 2$, since

$$|1/\zeta(s)| = |\prod_p (1-p^{-s})| \leqslant \prod_p (1+p^{-\sigma}) < \zeta(\sigma) \leqslant \zeta(2)$$

when $\sigma \geqslant 2$. If $1 < \sigma \leqslant 2$, $t \geqslant 2$, then by (11)

$$(\sigma-1)^3 \leqslant \{(\sigma-1)\zeta(\sigma)\}^3 |\zeta(\sigma+ti)|^4 |\zeta(\sigma+2ti)|$$
$$\leqslant A_1^3 . |\zeta(\sigma+ti)|^4 . A_2 \log(2t),$$

by (4). Hence (since $\log(2t) \leqslant \log t^2 = 2\log t$)

$$(12) \qquad |\zeta(\sigma+ti)| \geqslant \frac{(\sigma-1)^{\frac{3}{4}}}{A_3(\log t)^{\frac{1}{4}}} \qquad (1 \leqslant \sigma \leqslant 2,\ t \geqslant 2)$$

(the inequality being trivial for $\sigma = 1$). Now let $1 < \eta < 2$. Then, if $1 \leqslant \sigma \leqslant \eta$, $t \geqslant 2$,

$$|\zeta(\sigma+ti) - \zeta(\eta+ti)| = \left|\int_\sigma^\eta \zeta'(u+ti)\,du\right| \leqslant A_4 \log^2 t . (\eta-1)$$

[1] $f^n(s)$ is the nth power, $f^{(n)}(s)$ the nth derivative, of $f(s)$.

by (5). Hence

$$|\zeta(\sigma+ti)| \geqslant |\zeta(\eta+ti)| - A_4(\eta-1)(\log t)^2$$
$$> \frac{(\eta-1)^{\frac{3}{4}}}{A_3(\log t)^{\frac{1}{4}}} - A_4(\eta-1)(\log t)^2$$

by (12). The last inequality holds also, in virtue of (12), for $\eta \leqslant \sigma \leqslant 2$, $t \geqslant 2$; it is therefore true for $1 < \sigma \leqslant 2$, $t \geqslant 2$. Now choose $\eta = \eta(t)$ so that

$$\frac{(\eta-1)^{\frac{3}{4}}}{A_3(\log t)^{\frac{1}{4}}} = 2A_4(\eta-1)(\log t)^2,$$

i.e.
$$\eta = 1 + (2A_3A_4)^{-4}(\log t)^{-9},$$

assuming t large enough (say $t > t_0$) to ensure that $\eta < 2$. Then

$$|\zeta(\sigma+ti)| \geqslant A_4(\eta-1)(\log t)^2 = A_5(\log t)^{-7}$$

for $1 < \sigma \leqslant 2$, $t > t_0$. This proves (9) (with $A = 7$).

The inequality (10) is only one example of an infinite number of inequalities of the type

$$c_0 + c_1\cos\theta + \ldots + c_m\cos m\theta \geqslant 0,$$

which might be used for the same purpose. The essential requirements are that the coefficients c should be positive and that $c_1 > c_0$. Another example is

$$5 + 8\cos\theta + 4\cos 2\theta + \cos 3\theta \geqslant 0.$$

The proof which we have given (of the first part of Theorem 10) is, in essentials, that of Hadamard.[1] A quite different proof was given by de la Vallée Poussin[2]; his proof is more elaborate and demands a much more profound knowledge of the properties of $\zeta(s)$.

5. Fundamental formula. Having established the requisite properties of $\zeta(s)$ we now seek to deduce from them a property of $\psi(x)$. The fundamental relation [(17), p. 18] between these two functions expresses $f(s) = -\zeta'(s)/\zeta(s)$ in terms of $\psi(x)$. For our present purpose it is evidently desirable to have a relation in the opposite sense—one expressing $\psi(x)$ in terms of $f(s)$. The direct discussion of $\psi(x)$ introduces, however, delicate questions of convergence, and to avoid these we shall work in the first instance with the function

$$\text{(13)} \quad \psi_1(x) = \int_0^x \psi(u)\,du = \int_1^x \psi(u)\,du = \sum_{n \leqslant x} (x-n)\Lambda(n).$$

(The identity of the last two expressions is a simple consequence

[1] Hadamard 2. See also Mertens 2; H, i, 242–258.

[2] De la Vallée Poussin 1.

of Theorem A, p. 18.) The transition from $\psi_1(x)$ to $\psi(x)$ will be found to be a comparatively simple matter. We shall first prove the *fundamental formula*

$$(14)\quad \psi_1(x)=\frac{1}{2\pi i}\int_{c-\infty i}^{c+\infty i}\frac{x^{s+1}}{s(s+1)}\left(-\frac{\zeta'(s)}{\zeta(s)}\right)ds \qquad (x>0,\ c>1),$$

where the path of integration is the straight line $\sigma=c$. The proof is based on

Theorem B. *If k is a positive integer, $c>0$, $y>0$, then*

$$\frac{1}{2\pi i}\int_{c-\infty i}^{c+\infty i}\frac{y^s\,ds}{s(s+1)\dots(s+k)}=\begin{cases}0 & (y\leqslant 1),\\ \dfrac{1}{k!}\left(1-\dfrac{1}{y}\right)^k & (y\geqslant 1).\end{cases}$$

The integral is absolutely convergent, since the integrand has modulus less than $y^c\,|t|^{-k-1}$ on the line of integration, and $k>0$. Denote by J the infinite integral and by J_T the integral from $c-Ti$ to $c+Ti$ (each with the factor $1/2\pi i$). We use Cauchy's theorem of residues to replace the line of integration in J_T by an arc of the circle C having its centre at $s=0$ and passing through the points $s=c\pm Ti$. If $y\geqslant 1$, we use the arc C_1 which lies to the left of the line $\sigma=c$, assuming T so large that $R>2k$, where R is the radius of C. This gives

$$(15)\qquad J_T=S+J(C_1),$$

where S is the sum of the residues of the integrand at its poles $s=0,\ -1,\ \dots,\ -k$, and $J(C_1)$ is the integral along C_1. Now on C_1 we have $\sigma\leqslant c$, and so $|y^s|=y^\sigma\leqslant y^c$, since $y\geqslant 1$; also

$$|s+n|\geqslant R-k>\tfrac{1}{2}R \qquad (n=0,\ 1,\ \dots,\ k).$$

It follows that

$$|J(C_1)|<\frac{1}{2\pi}\cdot\frac{y^c}{(\frac{1}{2}R)^{k+1}}\cdot 2\pi R=\frac{2^{k+1}y^c}{R^k}<\frac{2^{k+1}y^c}{T^k}.$$

Hence, by (15), $J_T\to S$ as $T\to\infty$; thus $J=S$. But

$$S=\sum_{r=0}^{k}\frac{y^{-r}}{(-1)^r r!\,(k-r)!}=\frac{1}{k!}(1-y^{-1})^k,$$

which proves the theorem when $y\geqslant 1$. The proof is similar in the case $y\leqslant 1$, except that the right-hand arc C_2 of C is used and no poles are passed over.

To deduce (14) we have, for $x > 0$,

$$\frac{\psi_1(x)}{x} = \sum_{n \leqslant x} \left(1 - \frac{n}{x}\right) \Lambda(n) = \sum_{n=1}^{\infty} \frac{\Lambda(n)}{2\pi i} \int_{c-\infty i}^{c+\infty i} \frac{(x/n)^s}{s(s+1)} ds,$$

if $c > 0$, by Theorem B. If $c > 1$, the order of summation and integration may be interchanged, since

$$\sum_1^{\infty} \int_{c-\infty i}^{c+\infty i} \left| \frac{\Lambda(n)(x/n)^s}{s(s+1)} ds \right| < x^c \sum_1^{\infty} \frac{\Lambda(n)}{n^c} \int_{-\infty}^{\infty} \frac{dt}{c^2 + t^2}$$

is finite.[1] Hence

$$\frac{\psi_1(x)}{x} = \frac{1}{2\pi i} \int_{c-\infty i}^{c+\infty i} \frac{x^s}{s(s+1)} \sum_{n=1}^{\infty} \frac{\Lambda(n)}{n^s} ds$$

$$= \frac{1}{2\pi i} \int_{c-\infty i}^{c+\infty i} \frac{x^s}{s(s+1)} \left(-\frac{\zeta'(s)}{\zeta(s)}\right) ds,$$

which is equivalent to (14).

If we integrate by parts in (17), p. 18, we obtain

$$\text{(16)} \qquad \frac{1}{s(s+1)} \left(-\frac{\zeta'(s)}{\zeta(s)}\right) = \int_0^{\infty} \frac{\psi_1(x)}{x^{s+2}} dx \qquad (\sigma > 1)$$

(since $\psi_1(x) = 0$ for $x < 2$). Equations (14) and (16) may be regarded as an instance of Mellin's inversion formula; or by a change of variable they may be exhibited as a pair of reciprocal formulae of the type occurring in the theory of 'Fourier transforms', and either may be deduced from the other by general theorems. These formulae are special cases of general formulae in the theory of Dirichlet's series, valid under much wider conditions.[2]

6. We are now in a position to obtain an asymptotic formula for $\psi_1(x)$. This is the crucial step in the proof of the prime number theorem.

Theorem 11. *We have*

$$\psi_1(x) \sim \tfrac{1}{2} x^2$$

when $x \to \infty$.

[1] The inversion depends on the general principle (to which we shall constantly appeal) that, if S_x and T_y denote summations or integrations with respect to the variables indicated, then

$$S_x T_y f(x, y) = T_y S_x f(x, y),$$

provided that one side is finite when f is replaced by $|f|$, and that appropriate integrability conditions (always fulfilled in our applications) are satisfied when integrals are involved. For the special case $f \geqslant 0$ see de la Vallée Poussin's *Cours d'Analyse* (4th or 5th ed.), ii, §§ 14, 17. The general case may be reduced to this by the familiar device of writing $f = u + vi$, $u = u_1 - u_2$, $|u| = u_1 + u_2$, $v = v_1 - v_2$, $|v| = v_1 + v_2$, and applying the special case to u_1, u_2, v_1, v_2 separately.

[2] HR, Theorems 13, 24, 29, 39, 40.

Suppose throughout $x > 1$. By (14)

$$\frac{\psi_1(x)}{x^2} = \int_{(c)} g(s)\, x^{s-1}\, ds, \tag{17}$$

where $c > 1$, (c) denotes the line $\sigma = c$, and

$$g(s) = \frac{1}{2\pi i}\frac{1}{s(s+1)}\left(-\frac{\zeta'(s)}{\zeta(s)}\right) = -\frac{1}{2\pi i}\cdot\frac{1}{s(s+1)}\cdot\zeta'(s)\cdot\frac{1}{\zeta(s)}.$$

By Theorems 8, 9, 10, $g(s)$ is regular in $\sigma \geq 1$, except at $s = 1$, and

$$|g(s)| < A_1 . |t|^{-2} . A_2 (\log|t|)^2 . A_3 (\log|t|)^{A_4}$$

$$< |t|^{-\frac{3}{2}} \quad (\sigma \geq 1,\ |t| > t_0). \tag{18}$$

Take $\epsilon > 0$, and let $L = L(\epsilon)$ be the infinite broken line

$$L_1 + L_2 + L_3 + L_4 + L_5$$

shown in the figure, where $T = T(\epsilon)$ is chosen so that

$$\int_T^\infty |g(1+ti)|\, dt < \epsilon, \tag{19}$$

and then $\alpha = \alpha(T) = \alpha(\epsilon)$ $(0 < \alpha < 1)$ so that the rectangle $\alpha \leq \sigma \leq 1$, $-T \leq t \leq T$ contains no zero of $\zeta(s)$. The first choice is possible by (18), and the second because $\zeta(s)$ has no zeros on the line $\sigma = 1$ (by Theorem 10) and (as a regular function) at most a finite number in the region $\frac{1}{2} \leq \sigma < 1$, $-T \leq t \leq T$. Applying Cauchy's theorem to (17), we obtain

$$\frac{\psi_1(x)}{x^2} = \tfrac{1}{2} + \int_L g(s)\, x^{s-1}\, ds = \tfrac{1}{2} + J, \tag{20}$$

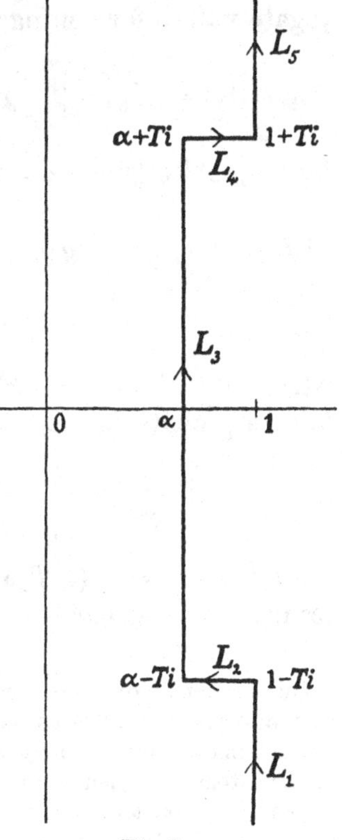

Fig. 1

say, the term $\frac{1}{2}$ arising from the pole at $s = 1$ (at which point $\zeta'(s)/\zeta(s)$, as the logarithmic derivative of a function having a pole of order 1, has a simple pole with residue -1). For by our choice of L the integrand is regular between and on the lines (c) and L (except at $s = 1$), and, if we first integrate round a closed contour bounded by portions of (c) and L and by segments of the lines $t = \pm U$, where $U > \text{Max}(t_0, T)$, the integrals along the latter segments are in absolute value

$$< (c-1) \max_{1 \leqslant \sigma \leqslant c} |g(\sigma \pm Ui)| x^{\sigma-1} < (c-1) U^{-\frac{3}{2}} x^{c-1}$$

by (18), and therefore tend to 0 when $U \to \infty$; (18) shows also that J is absolutely convergent.

Write $J = J_1 + J_2 + J_3 + J_4 + J_5$, where $J_1, \ldots, J_5$ are the integrals along $L_1, \ldots, L_5$, respectively. Since $g(s)x^{s-1}$ takes conjugate values for conjugate values of s,

$$|J_1| = |J_5| = \left| \int_T^\infty g(1+ti) x^{ti} dt \right| \leqslant \int_T^\infty |g(1+ti)| \, dt < \epsilon$$

by (19). Also (since $x > 1$)

$$|J_2| = |J_4| = \left| \int_\alpha^1 g(\sigma + Ti) x^{\sigma + Ti - 1} d\sigma \right| \leqslant M \int_\alpha^1 x^{\sigma-1} d\sigma < \frac{M}{\log x},$$

$$|J_3| \leqslant M x^{\alpha-1} . 2T,$$

where $M = M(T, \alpha) = M(\epsilon)$ is the maximum of $|g(s)|$ on the finite segments L_2, L_3, L_4. Hence, by (20),

$$\left| \frac{\psi_1(x)}{x^2} - \frac{1}{2} \right| = |J| < 2\epsilon + \frac{2M}{\log x} + \frac{2MT}{x^{1-\alpha}} < 3\epsilon$$

for all $x > x_0 = x_0(\epsilon, T, \alpha, M) = x_0(\epsilon)$. This proves the desired result, that $\psi_1(x)/x^2 \to \frac{1}{2}$ as $x \to \infty$.

Our object in moving the line of integration to the left is, of course, to make $|x^{s-1}| = x^{\sigma-1}$ as small as possible on the new contour; but limitations are imposed by the possibility of encountering singularities of the integrand arising from zeros of $\zeta(s)$. There is an alternative proof in which no part of the new line of integration lies to the left of the line $\sigma = 1$. We have in fact, by (14) and Theorem B (with a change of variable $s = s' + 1$),

$$\frac{\psi_1(x)}{x^2} - \frac{1}{2}\left(1 - \frac{1}{x}\right)^2 = \frac{1}{2\pi i} \int_{(c)} \frac{x^{s-1}}{s(s+1)} \left(-\frac{\zeta'(s)}{\zeta(s)} - \frac{1}{s-1} \right) ds = \frac{1}{2\pi i} \int_{(c)} h(s) x^{s-1} ds,$$

say, if $x > 1$, $c > 1$. We move the path of integration to the left as before, but in this case we may take the line $\sigma = 1$ as the new path, since $h(s)$ is regular at $s = 1$. Hence

$$\frac{\psi_1(x)}{x^2} - \frac{1}{2}\left(1 - \frac{1}{x}\right)^2 = \frac{1}{2\pi}\int_{-\infty}^{\infty} h(1 + ti)\, e^{\xi ti}\, dt, \tag{21}$$

where $\xi = \log x$. Since

$$\int_{-\infty}^{\infty} |h(1 + ti)|\, dt$$

is convergent, it follows from a simple extension of the 'Riemann-Lebesgue theorem' in the theory of Fourier series that the integral on the right of (21) tends to 0 when $\xi \to \infty$. This proves the theorem.

7. We have now to deduce from Theorem 11 the corresponding relation for $\psi(x)$. The deduction is based on

Theorem C. *Let $c_1, c_2, \ldots$ be a given sequence of numbers, and let*

$$C(x) = \sum_{n \leqslant x} c_n, \quad C_1(x) = \int_0^x C(u)\, du = \sum_{n \leqslant x} (x - n)\, c_n.$$

If $c_n \geqslant 0$ $(n = 1, 2, \ldots)$, and $C_1(x) \sim Cx^c$ (as $x \to \infty$), where C and c are positive constants, then $C(x) \sim Ccx^{c-1}$.

(The identity of the two expressions for $C_1(x)$ is, again, a consequence of Theorem A.)

Let $0 < \alpha < 1 < \beta$. Since $c_n \geqslant 0$, $C(u)$ is an increasing function (in the wide sense); hence, for $x > 0$,

$$C(x) \leqslant \frac{1}{\beta x - x}\int_x^{\beta x} C(u)\, du = \frac{C_1(\beta x) - C_1(x)}{(\beta - 1)x},$$

$$\frac{C(x)}{x^{c-1}} \leqslant \frac{1}{\beta - 1}\left(\frac{C_1(\beta x)}{(\beta x)^c}\beta^c - \frac{C_1(x)}{x^c}\right).$$

Let $x \to \infty$, keeping β fixed; then, since $C_1(y)/y^c \to C$ when $y \to \infty$,

$$\overline{\lim}\, \frac{C(x)}{x^{c-1}} \leqslant C\frac{\beta^c - 1}{\beta - 1}.$$

By considering the interval $(\alpha x, x)$ we prove similarly that

$$\underline{\lim}\, \frac{C(x)}{x^{c-1}} \geqslant C\frac{1 - \alpha^c}{1 - \alpha}.$$

By taking α and β near enough to 1 we can make the two expressions on the right as near as we please to Cc (the derivative of Cx^c for $x = 1$). Hence

$$\overline{\lim}\, C(x)/x^{c-1} = \underline{\lim}\, C(x)/x^{c-1} = Cc.$$

The inference from '$C(x) \sim Ccx^{c-1}$' to '$C_1(x) \sim Cx^c$' is, of course, valid (and almost trivial) without the restriction $c_n > 0$. That the converse is not true unconditionally may be seen from the example $c_n = 1 + (-1)^n n$. For some general observations on Theorem C, see § 10 below.

8. The prime number theorem. The proof of the prime number theorem now consists merely in assembling the results of the preceding theorems.

Theorem 12. *We have*

$$\pi(x) \sim \frac{x}{\log x}$$

when $x \to \infty$.

By Theorem 11, $\psi_1(x) \sim \frac{1}{2}x^2$. Since $\Lambda(n) > 0$, it follows from Theorem C that $\psi(x) \sim x$. This, by Theorem 3, is equivalent to $\pi(x) \sim x/\log x$.

For the nth prime p_n, Theorem 12 gives the following result.

Theorem 13. *We have*

$$p_n \sim n \log n$$

when $n \to \infty$.

For, if $y = \pi(x)$, then, when $x \to \infty$,

$$\frac{y \log x}{x} \to 1, \tag{22}$$

so that
$$\log y + \log\log x - \log x \to 0.$$

Hence
$$\frac{\log y}{\log x} \to 1.$$

This combined with (22) gives

$$\frac{y \log y}{x} \to 1,$$

whence the result follows on taking $x = p_n$, since $\pi(p_n) = n$.

Conversely, Theorem 12 can be deduced from Theorem 13, so that the two theorems are equivalent. For suppose Theorem 13 true, and let $n = n(x)$ be defined by $p_n \leqq x < p_{n+1}$. Then $p_n \sim n \log n$ and $p_{n+1} \sim (n+1)\log(n+1) \sim n \log n$ when $x \to \infty$, whence it follows that $x \sim n \log n$, or $x \sim y \log y$, since $y = \pi(x) = n$. From this we deduce, as above, that $\log x \sim \log y$, and thence that $y \sim x/\log x$.

9. The above proof of the prime number theorem is based on the fact that $\zeta(s)$ has no zeros on the line $\sigma = 1$, and it is natural to ask whether the truth of the theorem depends essentially on this property of $\zeta(s)$. We shall show that this is so, by proving that the property in question can be deduced at once from the prime number theorem, if we take as known the properties of $\zeta(s)$ embodied in Theorem 8.

By equation (17), p. 18, we have, for $\sigma > 1$,

$$\int_1^\infty \frac{\psi(x) - x}{x^{s+1}} dx = -\frac{1}{s}\frac{\zeta'(s)}{\zeta(s)} - \frac{1}{s-1} = \phi(s),$$

say; $\phi(s)$ is regular in $\sigma > 0$ except (possibly) for simple poles at zeros of $\zeta(s)$. Now suppose the prime number theorem true, i.e. $\psi(x) = x + o(x)$. Then, given $\epsilon > 0$, we have $|\psi(x) - x| < \epsilon x$ for $x > x_0 = x_0(\epsilon)\ (> 1)$. Hence, for $\sigma > 1$,

$$|\phi(s)| < \int_1^{x_0} \frac{|\psi(x) - x|}{x^2} dx + \int_{x_0}^\infty \frac{\epsilon}{x^\sigma} dx < K + \frac{\epsilon}{\sigma - 1},$$

where $K = K(x_0) = K(\epsilon)$. Thus

$$|(\sigma - 1)\phi(\sigma + ti)| < K(\sigma - 1) + \epsilon < 2\epsilon$$

for $1 < \sigma < \sigma_0 = \sigma_0(\epsilon, K) = \sigma_0(\epsilon)$. Hence, for any fixed t,

$$(\sigma - 1)\phi(\sigma + ti) \to 0$$

as $\sigma \to 1 + 0$. This shows that the point $1 + ti$ cannot be a zero of $\zeta(s)$, for in that case $(\sigma - 1)\phi(\sigma + ti)$ would tend to a limit different from 0, namely the residue of $\phi(s)$ at the simple pole $1 + ti$.

10. As we stated in the Introduction a 'real variable' proof of the prime number theorem, that is to say a proof not involving explicitly or implicitly the notion of an analytic function of a complex variable, has never been discovered, and we can now understand why this should be so. For the preceding discussion shows that the theorem is inseparably bound up with the behaviour of $\zeta(s)$ on the whole of the line $\sigma = 1$.

The argument used in the proof of Theorem 6 will prove generally that, if

$$f(s) = \sum_1^\infty \frac{c_n}{n^s} \quad (s > 1), \qquad C(x) = \sum_{n \leqslant x} c_n,$$

then

(23) $\quad$ '$C(x)/x \to C$' implies '$(s-1)f(s) \to C$'

(when $x \to \infty$ and $s \to 1 + 0$ respectively). If the converse of this were true, the prime number theorem would follow at once on taking $c_n = \Lambda(n)$. But the converse is false, and the discussion of § 9 helps us to construct an example to show this. We have only to take an $f(s)$

with properties rather like those of $-\zeta'(s)/\zeta(s)$ but having singularities at points of the line $\sigma = 1$ other than $s = 1$, for example

$$(24)\quad f(s) = \zeta(s) + \tfrac{1}{2}\zeta(s-i) + \tfrac{1}{2}\zeta(s+i) = \sum_1^\infty \frac{1+\cos(\log n)}{n^s} = \sum_1^\infty \frac{c_n}{n^s}.$$

In this case $(s-1)f(s) \to 1$ when $s \to 1+0$. But, as $f(s)$ has poles at $s = 1 \pm i$, the argument of § 9 shows that $C(x)/x$ cannot tend to 1 (nor therefore, by (23), to any limit at all) when $x \to \infty$. And, indeed, it is easily verified directly (e.g. by Theorem A) that

$$C(x) = x + \frac{x}{\sqrt{2}} \cos\left(\log x - \frac{\pi}{4}\right) + o(x).$$

Our theorems and their mutual relationships are best looked at from the point of view of 'Abelian' and 'Tauberian' theorems. The inference from '$C(x) \sim Ccx^{c-1}$' to '$C_1(x) \sim Cx^c$' in § 7 (p. 36) is an Abelian theorem—an inference, of general validity, from the behaviour of a function to that of an average; Theorem C, a 'converse' valid only in virtue of a definite restriction on the function averaged, is a Tauberian theorem. By using other Tauberian theorems instead of Theorem C, and working with the appropriate auxiliary functions in place of $\psi_1(x)$, we can give to the above proof of the prime number theorem a great variety of forms without change of principle.[1]

Another example of an Abelian theorem is provided by (23), $f(s)$ being expressed as an average of $C(x)$ by the formula analogous to (17), p. 18. The converse of this is, however, not true even with the restriction $c_n \geqslant 0$, as the example (24) shows. There are indeed Tauberian theorems for Dirichlet's series and power series with positive coefficients (due to Hardy and Littlewood), which may be adopted as a basis for a proof of the prime number theorem. But these theorems have to be applied to the auxiliary function $\sum \Lambda(n) z^n$ $(0 < z < 1)$, whose behaviour is investigated by complex variable methods[2]; applied directly to $\sum \Lambda(n) n^{-s}$ they yield only a proof of the relation

$$\sum_{n \leqslant x} \frac{\Lambda(n)}{n} \sim \log x,$$

already obtained in an elementary way in Chapter I (Theorem 7).

11. The investigations of this chapter establish a mutual relationship between the prime number theorem and the theorem that $\zeta(s)$ has no zeros on the line $\sigma = 1$. But we cannot infer from them the *equivalence*

[1] The proof given in the text is a slight modification of one due to Landau. For other proofs see **H**, **BC**, and the references there given.

[2] Hardy and Littlewood **1**, **2** (127–134). This proof has now been placed on a more elementary basis through the discovery by Karamata (**1**) of a simple proof of the relevant Tauberian theorem. In principle Karamata's proof is similar to the proof of the fundamental theorem of Wiener's general Tauberian theory referred to in § 11, but the particular case happens to admit of special treatment in a remarkably simple way (cf. Wiener **2** (51)).

in any sense of these two propositions, since we have used in our proof of the prime number theorem a subsidiary theorem on the order of magnitude of $\zeta'(s)/\zeta(s)$. This defect has been removed by N. Wiener, who has succeeded in proving the prime number theorem without using (in addition to Theorem 8) any property of $\zeta(s)$ beyond that of having no zeros on the line $\sigma = 1$. Wiener's work on this subject is an application of his general theory of Tauberian theorems.[1]

12. The fact that the prime number theorem cannot at present be proved by real variable methods has given rise to a convenient classification of theorems according to their 'depth'. We call a theorem 'elementary' or 'transcendental' according as it can or cannot be proved without the theory of functions of a complex variable; and two theorems are called 'equivalent' if they can be derived from one another by 'elementary' methods.

One of the most striking instances of 'equivalent' theorems arises out of the equation

$$(25)\quad 0 = 1 - \frac{1}{2} - \frac{1}{3} - \frac{1}{5} + \frac{1}{6} - \frac{1}{7} + \frac{1}{10} - \frac{1}{11} - \frac{1}{13} + \frac{1}{14} + \frac{1}{15} - \dots$$

given by Euler in 1748.[2] The terms are the reciprocals of the 'square-free' numbers q (positive integers containing no repeated prime factor) preceded by the sign + or − according as q has an even or an odd number of prime factors; in modern notation the equation is equivalent to $\Sigma\mu(n)/n = 0$, where $\mu(n)$ is Möbius's function. Euler's purely formal argument, which amounts effectively to saying that the right-hand side is

$$\Pi(1 - p^{-1}) = 1/\zeta(1) = 1/\infty = 0,$$

can be modified (by introducing a variable s and making $s \to 1 + 0$) and made to show that the series, if convergent, must have the sum 0. But the convergence of the series was first established by von Mangoldt in 1897, and his proof was based on the detailed theory of $\zeta(s)$.[3] It is now known that the convergence of this series is 'equivalent' to the prime number theorem.[4] Thus Euler's equation (25), if interpreted as implying the convergence of the series on the right, is a 'transcen-

[1] Wiener **1, 2**. Wiener's original discussion was based on a Tauberian theorem for Lambert's series, of which the prime number theorem was already known to be a comparatively simple consequence. This theorem and its relation to the theory of numbers were first investigated by Hardy and Littlewood (**3**), but their proof does not provide a proof of the prime number theorem, since it depends explicitly on a theorem a little deeper than the prime number theorem itself. Wiener's contribution consisted in obtaining an independent proof of the Tauberian theorem for Lambert's series, the proof that the conditions of his general theorem are satisfied in this case being based on the fact that $\zeta(1 + ti) \neq 0$ for real t. Alternative proofs of the prime number theorem depending on Wiener's theory but not using Lambert's series were subsequently found by Ikehara (**1**) and by Wiener himself (**2**).

[2] Euler **2**, § 227, Exemplum I.

[3] von Mangoldt **3**; H, ii, 567–616.

[4] Landau, **3**.

dental' theorem. It is interesting to note that, although the series is the formal expansion of $\Pi(1-p^{-1})$, it is an 'elementary' theorem that the product of the first n factors of the infinite product tends to 0 as $n \to \infty$ (Theorem 1), but a 'transcendental' theorem that the sum of the first n terms of the series tends to a limit.

The distinction between 'elementary' and 'transcendental' is, of course, relative to the existing state of knowledge, and is in any case rather indefinite, since an argument which makes no explicit mention of analytic functions may nevertheless involve closely related ideas. Thus Wiener's theory referred to in § 11 enables us to deduce the prime number theorem from the properties of $\zeta(s)$ without direct appeal to complex variable theory. But Wiener's work is based on the theory of Fourier transforms, a theory of about the same degree of sophistication as the theory of functions of a complex variable, and not unrelated to it; and for this reason Wiener himself does not regard his theory as strictly 'elementary' in the sense explained above.

CHAPTER III

FURTHER THEORY OF $\zeta(s)$. APPLICATIONS

1. In this chapter we shall develop the theory of $\zeta(s)$ systematically, and apply our results to the further study of $\psi(x)$ and $\pi(x)$.

2. Analytic continuation and functional equation. We resume the question of analytic continuation, and we now seek to determine the whole domain of existence of $\zeta(s)$. One method is to extend the domain step by step by repeated partial integrations in the formula (3), p. 26, or, what is effectively equivalent, by the application of the Euler-Maclaurin sum-formula. But we prefer to adopt other methods, which have the advantage of leading naturally to an important functional relation connecting $\zeta(s)$ with $\zeta(1-s)$.

Theorem 14. *The function* $\zeta(s)$, *defined by* $\zeta(s)=\Sigma n^{-s}$ *when* $\sigma>1$, *exists as a single-valued analytic function, having as its only singularity in the finite part of the plane a simple pole at* $s=1$, *with residue* 1.

Further, $\zeta(s)$ *satisfies the functional equation*

$$\zeta(1-s)=2(2\pi)^{-s}\cos\tfrac{1}{2}\pi s\,\Gamma(s)\,\zeta(s). \tag{1}$$

Since $\Gamma(s)=\int_0^\infty y^{s-1}e^{-y}\,dy=n^s\int_0^\infty x^{s-1}e^{-nx}\,dx \quad (\sigma>0, n>0)$,
we have, for $\sigma>1$,

$$\Gamma(s)\,\zeta(s)=\sum_1^\infty\Gamma(s)\,n^{-s}=\sum_1^\infty\int_0^\infty x^{s-1}e^{-nx}\,dx=\int_0^\infty\frac{x^{s-1}}{e^x-1}\,dx, \tag{2}$$

on inverting summation and integration, and summing the geometric series. The inversion is legitimate (p. 32, footnote [1]), since

$$\sum_1^\infty\int_0^\infty|x^{s-1}e^{-nx}|\,dx=\sum_1^\infty\int_0^\infty x^{\sigma-1}e^{-nx}\,dx=\sum_1^\infty\Gamma(\sigma)\,n^{-\sigma}=\Gamma(\sigma)\,\zeta(\sigma)$$

is finite.

Now consider the integral

$$I(s)=\frac{1}{2\pi i}\int_C\frac{z^{s-1}}{e^{-z}-1}\,dz$$

taken along the (infinite) contour $C = C_1 + C_2 + C_3$ in Fig. 2, where $c < 2\pi$, and C_1, C_3 lie along the lower and upper edges, respectively, of a 'cut' in the z-plane along the negative real

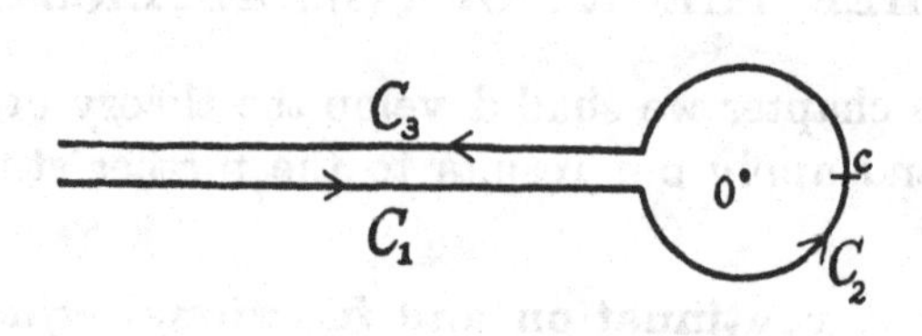

Fig. 2

axis. We define z^s, for any s, as $\exp(s \log z)$, where $\log z$ is continuous in the cut z-plane and real on the positive real axis. Then, if

$$z = re^{\theta i} \qquad (r > 0, -\pi < \theta < \pi),$$

we have

$$|z^s| = e^{\sigma \log r - t\theta} = r^\sigma e^{-t\theta}.$$

$I(s)$ (if convergent) is independent of c, by Cauchy's theorem, and is thus a function of s only. It is convergent for all s, and uniformly convergent in any fixed circle $|s| < \Delta$, since

$$\left|\frac{z^{s-1}}{e^{-z}-1}\right| = \frac{r^{\sigma-1}e^{\pm t\pi}}{e^r - 1} < \frac{r^{\Delta-1}e^{\Delta\pi}}{e^r-1} < e^{-\frac{1}{2}r}$$

on C_1 and C_3 if $r > r_0 = r_0(\Delta)$. Hence $I(s)$ is regular in $|s| < \Delta$ for any Δ, and is therefore an integral function. (See p. 25, footnote; $I(s)$ is reduced to integrals with respect to real variables by equation (3) below.)

Putting $z = re^{-\pi i}$, $ce^{\theta i}$, $re^{\pi i}$, on C_1, C_2, C_3 respectively, and writing $1/(e^{-z}-1) = g(z)$, we obtain

$$2\pi i\, I(s) = -\int_c^\infty r^{s-1}e^{-s\pi i} g(-r)\, dr$$

$$+\int_{-\pi}^{\pi} c^s e^{s\theta i} g(ce^{\theta i})\, i\, d\theta + \int_c^\infty r^{s-1}e^{s\pi i} g(-r)\, dr,$$

or

$$(3)\quad \pi I(s) = \sin s\pi \int_c^\infty r^{s-1} g(-r)\, dr + \frac{c^s}{2}\int_{-\pi}^{\pi} e^{s\theta i} g(ce^{\theta i})\, d\theta$$

$$= I_1(s, c) + I_2(s, c),$$

say. Since $zg(z)$ is regular in $|z|<2\pi$, we have $|zg(z)|\leqslant A_1$ for $|z|\leqslant\pi$; hence, if $c\leqslant\pi$,

$$(4)\qquad |I_2(s,c)|\leqslant\frac{c^\sigma}{2}\int_{-\pi}^{\pi}e^{-t\theta}\frac{A_1}{c}\,d\theta\leqslant\pi A_1e^{|t|\pi}c^{\sigma-1}.$$

Now suppose $\sigma>1$. Then, by (4), $I_2(s,c)\to 0$ as $c\to 0$ (s fixed). Hence

$$\pi I(s)=\lim_{c\to 0}I_1(s,c)=\sin s\pi\int_0^\infty r^{s-1}g(-r)\,dr=\sin s\pi\,\Gamma(s)\,\zeta(s)$$

by (2). Thus

$$(5)\qquad \zeta(s)=\frac{\pi I(s)}{\sin s\pi\,\Gamma(s)}=\Gamma(1-s)\,I(s).$$

Since $I(s)$ is an integral function, this equation, proved first for $\sigma>1$, defines $\zeta(s)$ as a meromorphic function whose poles are at those of the points $s=1, 2, 3, \ldots$ (the poles of $\Gamma(1-s)$) for which $I(s)\neq 0$. Now if s is an integer the integrand in $I(s)$ is a single-valued function of z, and $I(s)$ is its residue at $z=0$; hence $I(1)=-1$, $I(2)=I(3)=\ldots=0$. Thus the points $s=2, 3,\ldots$ are not poles of $\zeta(s)$ (as is otherwise obvious from the fact that they lie in the half-plane $\sigma>1$). And, since

$$(s-1)\,\zeta(s)=-\Gamma(2-s)\,I(s),$$

the point $s=1$ is a simple pole with residue $-\Gamma(1)\,I(1)=1$. This proves the first part of the theorem.

In proving the second part we shall suppose $\sigma<0$; this is permissible by the theory of analytic continuation, since each side of (1) is regular, except for poles, in the whole s-plane. Consider

$$I_N(s)=\frac{1}{2\pi i}\int_{C(N)}\frac{z^{s-1}}{e^{-z}-1}\,dz,$$

where $C(N)$ is the closed contour shown in Fig. 3, N being a positive integer. On the outer circle we have

$$z=Re^{\theta i}\qquad(-\pi\leqslant\theta\leqslant\pi),$$

$$\left|\frac{z^{s-1}}{e^{-z}-1}\right|=R^{\sigma-1}e^{-t\theta}\left|\frac{1}{e^{-z}-1}\right|<R^{\sigma-1}e^{|t|\pi}A_2.$$[1]

[1] $|1/(e^{-z}-1)|$ is bounded in the region $S(\delta)$ which remains when we remove from the z-plane the interiors of circles of radius $\delta\,(<\pi)$ with centres at $s=2n\pi i$ ($n=0,\pm 1,\pm 2,\ldots$); for it has period $2\pi i$, does not exceed $e/(e-1)$ for $|x|\geqslant 1$ ($z=x+yi$), and is continuous and therefore bounded for $-1\leqslant x\leqslant 1$, $-\pi\leqslant y\leqslant\pi$, $|z|\geqslant\delta$. And the circle $|z|=R$ lies entirely in $S(\frac{1}{2}\pi)$, for example.

Hence the contribution to $I_N(s)$ of this part of $C(N)$ is less than $R^\sigma e^{|t|\pi} A_2$ in modulus, and therefore tends to 0 when $N \to \infty$, since $\sigma < 0$. It follows that

$$I_N(s) \to I(s) \quad \text{as} \quad N \to \infty.$$

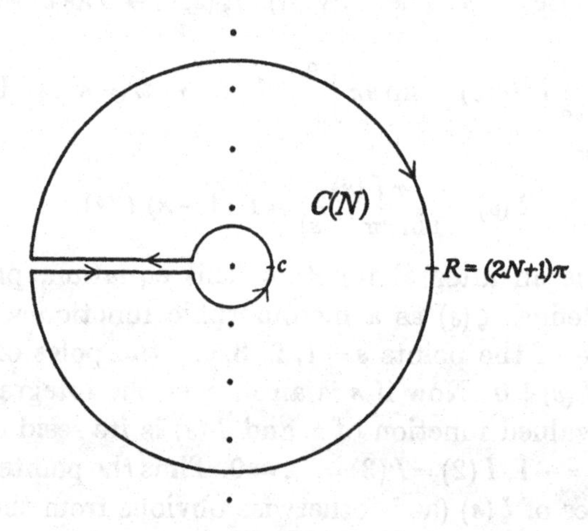

Fig. 3

Now by Cauchy's theorem of residues

$$I_N(s) = \sum_{n=1}^{N} \{(2n\pi i)^{s-1} + (-2n\pi i)^{s-1}\}$$

$$= \sum_{1}^{N} (2n\pi)^{s-1} (e^{\frac{1}{2}\pi(s-1)i} + e^{-\frac{1}{2}\pi(s-1)i})$$

$$= \sum_{1}^{N} (2n\pi)^{s-1} . 2\cos\tfrac{1}{2}\pi(s-1) = 2(2\pi)^{s-1} \sin\tfrac{1}{2}\pi s \sum_{1}^{N} n^{s-1}.$$

Hence

$$I(s) = \lim_{N\to\infty} I_N(s) = 2(2\pi)^{s-1} \sin\tfrac{1}{2}\pi s \sum_{1}^{\infty} n^{s-1}$$

$$= 2(2\pi)^{s-1} \sin\tfrac{1}{2}\pi s\, \zeta(1-s),$$

since $\Re(1-s) = 1-\sigma > 1$.

This combined with (5) gives

$$\zeta(s) = \frac{(2\pi)^s \sin\frac{1}{2}\pi s\, \zeta(1-s)}{\sin \pi s\, \Gamma(s)},$$

which is equivalent to (1).

3. The function $\xi(s)$. We can put the functional equation into a more symmetrical form by using familiar properties of $\Gamma(s)$.

Theorem 15. *The function*

$$\xi(s) = \tfrac{1}{2}s(s-1)\pi^{-\frac{1}{2}s}\Gamma(\tfrac{1}{2}s)\zeta(s) \tag{6}$$

is an integral function, and satisfies the functional equation

$$\xi(1-s) = \xi(s).$$

Further, $\xi(s)$ is real on the lines $t = 0$ and $\sigma = \frac{1}{2}$.

Also

$$\xi(0) = \xi(1) = \tfrac{1}{2}.$$

For the factor which multiplies $\zeta(s)$ in (1) is

$$2(2\pi)^{-s}\sin\tfrac{1}{2}\pi(1-s)\,.\,\Gamma(s)$$

$$= \frac{(2\pi)^{1-s}}{\Gamma(\frac{1}{2}-\frac{1}{2}s)\,\Gamma(\frac{1}{2}+\frac{1}{2}s)}\;\frac{2^{s-1}\Gamma(\frac{1}{2}s)\,\Gamma(\frac{1}{2}+\frac{1}{2}s)}{\pi^{\frac{1}{2}}}$$

$$= \frac{\pi^{-\frac{1}{2}s}\Gamma(\frac{1}{2}s)}{\pi^{-\frac{1}{2}(1-s)}\Gamma(\frac{1}{2}-\frac{1}{2}s)},$$

so that the function

$$\phi(s) = \pi^{-\frac{1}{2}s}\Gamma(\tfrac{1}{2}s)\zeta(s)$$

satisfies $\phi(1-s) = \phi(s)$. Now $\phi(s)$ is a meromorphic function whose poles lie symmetrically about the point $s = \frac{1}{2}$, since $\phi(1-s) = \phi(s)$. But the only pole in $\sigma > 0$ is a simple one at $s = 1$ (with residue $\pi^{-\frac{1}{2}}\Gamma(\frac{1}{2}) = 1$); hence the only poles in the whole s-plane are simple ones at $s = 0$ and $s = 1$. Thus

$$\xi(s) = \tfrac{1}{2}s(s-1)\phi(s)$$

is an integral function, and

$$\xi(1-s) = \tfrac{1}{2}(1-s)(-s)\phi(1-s) = \tfrac{1}{2}(s-1)s\phi(s) = \xi(s).$$

It is clear from (6), (5), (3), and the properties of $\Gamma(s)$ and of the various elementary functions involved, that $\xi(s)$ is real when s is real, and that $\xi(\sigma+ti)$ and $\xi(\sigma-ti)$ are conjugate.[1] Thus $\xi(\frac{1}{2}+ti)$ and $\xi(\frac{1}{2}-ti)$ are conjugate, and they are also equal by the functional equation; they are therefore real.

Finally

$$\xi(s) = \pi^{-\frac{1}{2}s}\Gamma(\tfrac{1}{2}s+1)\,.\,(s-1)\zeta(s),$$

[1] Alternatively, we may infer these results, in virtue of a known general theorem, from the fact that $\zeta(s)$ is a single-valued analytic function taking real values on the stretch $s > 1$ of the real axis.

so that $\xi(1)=\pi^{-\frac{1}{2}}\Gamma(\frac{3}{2}).1=\frac{1}{2}$; and $\xi(0)=\xi(1)$ by the functional equation.

Riemann wrote $\xi(t)$ for the function which we denote by $\xi(\frac{1}{2}+ti)$, and treated t as complex. The above notation is that adopted by Landau, who writes $\Xi(t)$ for Riemann's $\xi(t)$. The functional equation asserts that $\Xi(t)$ is an even function of t.

4. The above is one of the two proofs given by Riemann. The other, which gives directly the symmetrical form of the functional equation, is based on the identity

$$\sum_{-\infty}^{\infty} e^{-n^2\pi x}=\frac{1}{\sqrt{x}}\sum_{-\infty}^{\infty} e^{-n^2\pi/x} \qquad (x>0), \tag{7}$$

a special case of linear transformation of a theta-function.[1]

There is another type of proof based on 'Poisson's summation formula'

$$\sum_{-\infty}^{\infty} f(n)=\sum_{-\infty}^{\infty}\int_{-\infty}^{\infty} f(u)\cos 2\pi nu\,du. \tag{8}$$

Formally, this is the expansion, for $x=0$, of the periodic function

$$F(x)=\sum_{-\infty}^{\infty} f(n+x)$$

as a Fourier series $\Sigma(a_n\cos 2\pi nx+b_n\sin 2\pi nx)$. Sufficient conditions for its validity are that $f(x)$ and $f'(x)$ should be continuous for all x, except possibly for a finite number of values in the case of $f'(x)$, and that $f(x)$ and $|f'(x)|$ should be integrable over $(-\infty,\infty)$; for these conditions imply that $F(x)$ is continuous and of bounded variation in $0\leqq x\leqq 1$, and Jordan's criterion may be applied to $F(x)$.

The formula (8) may be used to prove (7), and this gives an indirect application to $\zeta(s)$. But (8) may be applied more directly in various ways, of which the following is an example.[2] Let

$$f(x)=f(x,s)=(x-\tfrac{1}{2})_{-s}-2x_{-s}+(x+\tfrac{1}{2})_{-s}, \quad \phi(s)=\sum_{-\infty}^{\infty} f(n,s),$$

where x_{-s} is x^{-s} if $x>0$ and 0 if $x\leqq 0$. For $\sigma>1$,

$$\begin{aligned}\phi(s)&=\sum_{1}^{\infty}(n-\tfrac{1}{2})^{-s}-2\sum_{1}^{\infty}n^{-s}+\sum_{0}^{\infty}(n+\tfrac{1}{2})^{-s}\\ &=2\sum_{1}^{\infty}2^s\{(2n-1)^{-s}+(2n)^{-s}\}-4\sum_{1}^{\infty}n^{-s}\\ &=2^{1+s}\zeta(s)-4\zeta(s)=4(2^{s-1}-1)\zeta(s).\end{aligned}$$

Now $\phi(s)$ is regular for $\sigma>-1$, since $f(n,s)=O(n^{-\sigma-2})$, uniformly in any bounded domain of s, as $n\to\infty$. Hence the equation

$$\zeta(s)=\tfrac{1}{4}(2^{s-1}-1)^{-1}\phi(s) \tag{9}$$

[1] Riemann 1; see also T, 43 (4), and the remark at the end of the last section. For the history of the functional equation see Landau 2.

[2] Mordell 1; cf. also Hardy 1. For references to other proofs see BC, 759–763; T 2, footnote.

gives the analytic continuation of $\zeta(s)$ in the half-plane $\sigma > -1$, as a regular function except possibly for poles at the points

$$s = 1 + 2n\pi i/\log 2 \qquad (n = 0, \pm 1, \pm 2, \ldots).$$

If s is real, and $-1 < s < 0$, $f(x)$ satisfies the conditions for (8), so that

$$\phi(s) = \sum_{-\infty}^{\infty} J(2\pi n, s) \qquad (-1 < s < 0), \tag{10}$$

where

$$J(\xi, s) = \int_{-\infty}^{\infty} \{(u - \tfrac{1}{2})_{-s} - 2u_{-s} + (u + \tfrac{1}{2})_{-s}\} \cos \xi u du.$$

The integrand in $J(\xi, s)$ is $O(u^{-\sigma-2})$ when $u \to \infty$, and $O(|u - u_0|^{-\sigma})$ when $u \to u_0$, for $u_0 = 0, \tfrac{1}{2}, -\tfrac{1}{2}$; and these relations hold uniformly for all real ξ and in any bounded domain of s. Hence $J(\xi, s)$ is continuous in s and ξ for $-1 < \sigma < 1$ and all real ξ, and regular in the strip $-1 < \sigma < 1$ for any fixed ξ. If $\xi \neq 0$ and $0 < s < 1$, we can split $J(\xi, s)$ into three integrals corresponding to the three terms in $\{\ldots\}$, these integrals being convergent at infinity by the second mean value theorem; making the substitutions $u = v + \tfrac{1}{2}$, v, $v - \tfrac{1}{2}$ in these integrals, and reuniting into a single integral, we obtain

$$\begin{aligned} J(\xi, s) &= \int_{-\infty}^{\infty} v_{-s} \{\cos \xi (v + \tfrac{1}{2}) - 2 \cos \xi v + \cos \xi (v - \tfrac{1}{2})\} dv \\ &= \int_{0}^{\infty} v^{-s} \cos \xi v (2 \cos \tfrac{1}{2}\xi - 2) dv \\ &= 2(\cos \tfrac{1}{2}\xi - 1) |\xi|^{s-1} \sin \tfrac{1}{2}\pi s \Gamma(1 - s), \end{aligned}$$

by a known formula for the gamma-function (easily deducible from Euler's integral by Cauchy's theorem). The final equation for $J(\xi, s)$, proved subject to $\xi \neq 0$, $0 < s < 1$, is valid for $-1 < \sigma < 1$ if $\xi \neq 0$, since each side is a regular function of s in this strip. And from the continuity in ξ we now deduce that $J(0, s) = 0$ in this strip (as may be verified directly). Substituting into (10) we obtain, for $-1 < s < 0$,

$$\begin{aligned} \phi(s) &= -8(2\pi)^{s-1} \sin \tfrac{1}{2}\pi s \Gamma(1 - s)(1 + 3^{s-1} + 5^{s-1} + \ldots) \\ &= 8(2\pi)^{s-1} \sin \tfrac{1}{2}\pi s \Gamma(1 - s)(2^{s-1} - 1)\zeta(1 - s), \end{aligned}$$

since $1 - s > 1$. Comparing with (9) we deduce

$$\zeta(s) = 2(2\pi)^{s-1} \sin \tfrac{1}{2}\pi s \Gamma(1 - s)\zeta(1 - s). \tag{11}$$

Since $\zeta(s)$ is regular, except for poles, in $\sigma > -1$, the right-hand side of (11) is regular, except for poles, in $\sigma < 2$. Thus (11) completes the continuation of $\zeta(s)$ over the whole plane (as a meromorphic function), and is valid for general values of s; we obtain (1) on changing s to $1 - s$. Now we have seen, from (9), that $\zeta(s)$ is certainly regular in $\sigma > 1$ and in $-1 < \sigma < 1$. Hence by (11) it is regular in $\sigma < 0$ and in $0 < \sigma < 2$, except possibly for a pole at $s = 1$ (arising from the factor $\Gamma(1 - s)$). These results together show that the point $s = 1$ is the only possible pole of $\zeta(s)$. With regard to this point we note first that $\phi(0) = 1$ (from the definition), so that $\zeta(0) = -\frac{1}{2}$ by (9). Hence by (11)

$$(s-1)\zeta(s) = -2(2\pi)^{s-1}\sin\tfrac{1}{2}\pi s\,\Gamma(2-s)\,\zeta(1-s) \to -2.1.1.\Gamma(1).(-\tfrac{1}{2}) = 1$$

as $s \to 1$, so that the point 1 is in fact a simple pole with residue 1.

Formally Poisson's formula transforms one side of (11) directly into the other, but some device is necessary to meet the difficulty that Σn^{-s} and Σn^{s-1} have no common domain of convergence.

5. Zeros. Before examining in detail the question of the existence of zeros of $\zeta(s)$ and $\xi(s)$, we state some important facts about the distribution of any zeros which may exist.

Theorem 16. (i) *The zeros of* $\xi(s)$ (*if any exist*) *are all situated in the strip* $0 < \sigma < 1$, *and lie symmetrically about the lines* $t = 0$ *and* $\sigma = \frac{1}{2}$.

(ii) *The zeros of* $\zeta(s)$ *are identical* (*in position and order of multiplicity*) *with those of* $\xi(s)$, *except that* $\zeta(s)$ *has a simple zero at each of the points* $s = -2, -4, -6, \ldots$.

(iii) $\xi(s)$ *has no zeros on the real axis.*

We have

$$\xi(s) = (s-1)\,\pi^{-\frac{1}{2}s}\,\Gamma(\tfrac{1}{2}s+1)\,\zeta(s) = h(s)\,\zeta(s)$$

say. Now $\zeta(s)$ has no zeros in $\sigma > 1$ (by Euler's product), and the same is true of $h(s)$. Hence $\xi(s)$ has no zero in $\sigma > 1$, and therefore none in $\sigma < 0$, since $\xi(s) = \xi(1-s)$. The zeros (if any) are symmetrical about the real axis, since $\xi(\sigma \pm ti)$ are conjugates, and about the point $s = \frac{1}{2}$ since $\xi(s) = \xi(1-s)$; they are therefore also symmetrical about the line $\sigma = \frac{1}{2}$. This proves (i).

The zeros of $\zeta(s)$ can differ from those of $\xi(s)$ only in so far as $h(s)$ has zeros or poles. The only zero of $h(s)$ is at $s = 1$, and this is not a zero of either function; for it is a pole of $\zeta(s)$, and $\xi(1) = \frac{1}{2}$ by Theorem 15. The poles of $h(s)$ are simple ones at $s = -2, -4, -6, \ldots$. Since these are points at which $\xi(s)$ is regular and not zero, they must be simple zeros of $\zeta(s)$. This proves (ii).

To prove (iii) it is sufficient, in virtue of (i) and (ii) and of the relations $\xi(0) = \xi(1) = \frac{1}{2}$, to show that $\zeta(s) \neq 0$ for $0 < s < 1$. Now

$$(1-2^{1-s})\,\zeta(s) = (1-2^{-s}) + (3^{-s} - 4^{-s}) + \ldots \qquad (\sigma > 0);$$

for this equation is evidently true when $\sigma > 1$, and, since

$$|(2n-1)^{-s} - (2n)^{-s}| = \left| s \int_{2n-1}^{2n} \frac{dx}{x^{s+1}} \right| < \frac{|s|}{(2n-1)^{\sigma+1}} < \frac{\Delta}{(2n-1)^{\delta+1}}$$

for $\sigma > \delta$, $|s| < \Delta$, where δ and Δ are fixed positive numbers, each side is regular in the half-plane $\sigma > 0$. When $0 < s < 1$ the equation gives $(1 - 2^{1-s})\zeta(s) > 0$, or $\zeta(s) < 0$.

The strip $0 \leqslant \sigma \leqslant 1$ is called the 'critical strip' and the line $\sigma = \frac{1}{2}$ the 'critical line'. The zeros $-2, -4, -6, \ldots$ of $\zeta(s)$, which are unimportant for most purposes, are called the 'trivial zeros'.

6. We have now to prove that $\xi(s)$ actually has zeros, or in other words that $\zeta(s)$ has zeros other than the trivial ones. We base our proof on the theory of integral functions. We shall give an account of the relevant parts of this theory in the next section, but we must first prove two theorems of general function-theory, which will find further applications in the course of this tract.

Theorem D. *Suppose that $f(z)$ is regular in the circle $|z - z_0| \leqslant R$, and has n zeros (at least) in $|z - z_0| \leqslant r\ (< R)$. Then, if $f(z_0) \neq 0$,*

$$\left(\frac{R}{r}\right)^n \leqslant \frac{M}{|f(z_0)|}, \tag{12}$$

where M is the maximum of $|f(z)|$ on $|z - z_0| = R$.

Multiple zeros are (as always) counted according to their order of multiplicity.

We may suppose $z_0 = 0$, since the general case reduces to this by the substitution $z = z_0 + z'$. Suppose $f(z)$ has zeros at the points $a_1, a_2, \ldots, a_n$ in $|z| \leqslant r$ (multiple zeros being allowed for by repetition). Then

$$f(z) = \phi(z) \prod_{\nu=1}^{n} \frac{R(z - a_\nu)}{R^2 - \bar{a}_\nu z},$$

where $\bar{a}_\nu$ is the conjugate of a_ν, and $\phi(z)$ is regular in $|z| \leqslant R$. On $|z| = R$ each factor of Π has modulus 1; hence

$$|\phi(z)| = |f(z)| \leqslant M \qquad (|z| = R).$$

Since $\phi(z)$ is regular in $|z| \leqslant R$, it follows (by the maximum modulus principle or by Cauchy's inequalities) that $|\phi(0)| \leqslant M$. Hence

$$|f(0)| = |\phi(0)| \prod_{\nu=1}^{n} \frac{|a_\nu|}{R} \leqslant M\left(\frac{r}{R}\right)^n,$$

which proves the theorem, since $f(0) \neq 0$.

Theorem D serves the same purpose, for many applications, as the more precise Jensen's formula, an identity from which the inequality (12) is an easy deduction.

Theorem E. *Suppose that, in* $|z - z_0| < R$,

$$f(z) = \sum_0^\infty c_n (z - z_0)^n$$

is regular and satisfies the (algebraic) inequality

$$\Re f(z) < U.$$

Then

$$|c_n| \leqslant \frac{2(U - \Re c_0)}{R^n} \qquad (n = 1, 2, 3, \ldots). \tag{13}$$

And, in $|z - z_0| \leqslant r < R$, *we have*

$$|f(z) - f(z_0)| \leqslant \frac{2r}{R - r}\{U - \Re f(z_0)\},$$

$$\left|\frac{f^{(\nu)}(z)}{\nu!}\right| \leqslant \frac{2R}{(R - r)^{\nu+1}}\{U - \Re f(z_0)\} \qquad (\nu = 1, 2, \ldots).$$

We may suppose again $z_0 = 0$. Let

$$\phi(z) = U - f(z) = U - c_0 - \sum_1^\infty c_n z^n = \sum_0^\infty b_n z^n \qquad (|z| < R),$$

and let C be the circle $|z| = r < R$. Then

$$b_n = \frac{1}{2\pi i}\int_C \frac{\phi(z)\,dz}{z^{n+1}} = \frac{1}{2\pi r^n}\int_{-\pi}^{\pi} (P + iQ)\, e^{-n\theta i}\, d\theta \qquad (n \geqslant 0), \tag{14}$$

where $\qquad \phi(re^{\theta i}) = P(r, \theta) + iQ(r, \theta) = P + iQ.$

By considering the integral of the regular function $\phi(z) z^{n-1}$ round the same circle we obtain

$$0 = \frac{r^n}{2\pi}\int_{-\pi}^{\pi} (P + iQ)\, e^{n\theta i}\, d\theta \qquad (n \geqslant 1).$$

Taking conjugates in this equation and combining with (14) we deduce

$$b_n r^n = \frac{1}{\pi}\int_{-\pi}^{\pi} Pe^{-n\theta i}\,d\theta \qquad (n \geqslant 1).$$

Now $P = U - \Re f(z) \geqslant 0$ in $|z| < R$ and in particular on C. Hence, if $n \geqslant 1$,

$$|b_n|\,r^n \leqslant \frac{1}{\pi}\int_{-\pi}^{\pi} |Pe^{-n\theta i}|\,d\theta = \frac{1}{\pi}\int_{-\pi}^{\pi} P\,d\theta = 2\Re b_0,$$

by (14). Making $r \to R$, we deduce

$$(15) \qquad |b_n|\,R^n \leqslant 2\beta_0 \qquad (n \geqslant 1),$$

where $\beta_0 = \Re b_0$; this is equivalent to (13) since $b_0 = U - c_0$ and $b_n = -c_n$ $(n \geqslant 1)$.

If $|z| \leqslant r < R$ we deduce from (15)

$$|\phi(z) - \phi(0)| = \left|\sum_1^\infty b_n z^n\right| \leqslant \sum_1^\infty 2\beta_0\left(\frac{r}{R}\right)^n = \frac{2\beta_0 r}{R-r},$$

and, if $\nu \geqslant 1$,

$$|\phi^{(\nu)}(z)| \leqslant \sum_{n=\nu}^\infty n(n-1)\dots(n-\nu+1)\frac{2\beta_0 r^{n-\nu}}{R^n}$$

$$= \left(\frac{d}{dr}\right)^\nu \sum_{n=0}^\infty 2\beta_0\left(\frac{r}{R}\right)^n = \left(\frac{d}{dr}\right)^\nu \frac{2\beta_0 R}{R-r} = \frac{2\beta_0 R\,.\,\nu!}{(R-r)^{\nu+1}}.$$

These inequalities are equivalent to those stated for $f(z)$, since

$$\phi(z) = U - f(z), \quad \beta_0 = U - \Re f(0).$$

Theorem E is due, in various stages of refinement, to Hadamard, Borel, and Carathéodory. In the form stated it is a 'best possible' theorem, as may be seen by considering the function $f(z) = z/(1+z)$, for which $\Re f(z) < \frac{1}{2}$ throughout $|z| < 1$.

7. Integral functions. Let $G(z)$ be an integral function; we suppose that $G(z)$ is not a constant, and that $G(0) \neq 0$.

If $G(z)$ has an infinity of zeros, these can be arranged in a sequence $a_1, a_2, a_3, \dots$ so that

$$0 < |a_1| \leqslant |a_2| \leqslant \dots, \quad |a_n| \to \infty;$$

zeros with the same modulus may be arranged among themselves in any order, and multiple zeros are allowed for by repetition. Consider the series $\Sigma|a_n|^{-\alpha}$. It is divergent for $\alpha = 0$, and, if it is convergent for some real value of α, it is convergent for every larger value, so that there exists (by a Dedekind section argument) a unique $\tau \geq 0$ such that the series is convergent for $\alpha > \tau$ and divergent for $\alpha < \tau$; for $\alpha = \tau$ it may converge or diverge. The number τ is called the *exponent of convergence* of the sequence $(|a_n|)$. Let $k+1$ be the least *integral* value of α for which the series is convergent; thus $k \geq 0$, and $k < \tau < k+1$ if τ is not an integer, while if τ is an integer $\tau = k$ or $k+1$ according as $\Sigma |a_n|^{-\tau}$ is divergent or convergent. Then

$$(16) \qquad G(z) = e^{H(z)} \prod_n \left\{ \left(1 - \frac{z}{a_n}\right) e^{\frac{z}{a_n} + \frac{1}{2}\left(\frac{z}{a_n}\right)^2 + \dots + \frac{1}{k}\left(\frac{z}{a_n}\right)^k} \right\},$$

where $H(z)$ is an integral function, and the product is absolutely convergent for all z (the exponent of e in the general factor to be interpreted as 0 if $k = 0$). This is Weierstrass's classical theorem on 'primary factors', which we take as known. We also assume, as part of Weierstrass's theory, the legitimacy of any formal transformations of (16) which may be made in the sequel.

If $G(z)$ has only a finite number of zeros (or none), we define $\tau = k = 0$. Then (16) is still valid, but Π is now a finite product (to be interpreted as 1 if there are no zeros).

If $H(z)$ (unique except for additive multiples of $2\pi i$) is a polynomial, we shall denote its degree by h.

We now define the *order* of an integral function. Let $M(r)$ be the maximum of $|G(z)|$ on $|z| = r$; since $G(z)$ is not a constant, $M(r)$ tends steadily to infinity with r (by the maximum modulus principle and Liouville's theorem). If the relation

$$(17) \qquad \log M(r) = O(r^\beta) \quad \text{as} \quad r \to \infty$$

holds for some real value of β, there exists a unique $\omega \geq 0$ such that it holds for every $\beta > \omega$ but for no $\beta < \omega$; for $\beta = \omega$ it may or may not hold. The number ω is called the order of $G(z)$. (The definition of ω has, of course, no reference to zeros, and does not presuppose, for example, the existence of the formula (16).)

We now establish a number of inequalities between τ, h, ω. These are to be understood in the sense that the existence of the smaller side is implied by that of the larger, which is presupposed in each case.

Theorem F 1. $\omega < \text{Max}(\tau, h)$.

Further, (17) *holds for* $\beta = \text{Max}(\tau, h)$, *if*

either (i) $\tau < h$,

or (ii) $\Sigma |a_n|^{-\tau}$ *is a convergent infinite series.*

We have, by (16),

$$G(z) = e^{H(z)} \prod_n \phi\left(\frac{z}{a_n}\right), \tag{18}$$

where $H(z)$ is a polynomial of degree h, and

$$\phi(\zeta) = (1-\zeta)\, e^{\zeta + \frac{\zeta^2}{2} + \dots + \frac{\zeta^k}{k}}.$$

If $|\zeta| < \frac{1}{2}$,

$$|\phi(\zeta)| = \left| e^{-\frac{\zeta^{k+1}}{k+1} - \frac{\zeta^{k+2}}{k+2} - \dots} \right| < e^{|\zeta|^{k+1}(1+\frac{1}{2}+\frac{1}{4}+\dots)} = e^{2|\zeta|^{k+1}} < e^{|2\zeta|^{k+1}}.$$

If $|\zeta| > \frac{1}{2}$,

$$|\phi(\zeta)| < (1+|\zeta|)\, e^{|\zeta| + \dots + |\zeta|^k} = (1+|\zeta|)\, e^{(|\zeta|^{1-k} + \dots + 1)|\zeta|^k}$$

$$< (1+|\zeta|)\, e^{(2^{k-1} + \dots + 1)|\zeta|^k} < e^{|2\zeta|^k + \log |3\zeta|}.$$

Hence, for all ζ,

$$|\phi(\zeta)| \leqslant e^{|2\zeta|^{\vartheta} + \vartheta^{-1}|3\zeta|^{\vartheta}}, \tag{19}$$

where ϑ is any number satisfying $k \leqslant \vartheta \leqslant k+1$, $\vartheta > 0$.

Choose $\vartheta > 0$ and such that $\Sigma |a_n|^{-\vartheta}$ (if an infinite series) is convergent. Such a choice is always compatible with the condition $k < \vartheta < k+1$, and ϑ may be taken arbitrarily near to τ; for if $\Sigma |a_n|^{-\tau}$ is a convergent infinite series (so that $\tau > 0$) we may take $\vartheta = \tau$, while in all other cases $k < \tau < k+1$ and we may take $\tau < \vartheta < k+1$. Then, by (19), we have, denoting by $K_1, K_2, \dots$ numbers depending only on ϑ and on the function G,

$$|\phi(\zeta)| < e^{K_1 |\zeta|^{\vartheta}},$$

and so, by (18),

$$|G(z)| < \exp\left(|H(z)| + K_1 \sum_n \left|\frac{z}{a_n}\right|^{\vartheta}\right),$$

$$M(r) < \exp(K_2 r^h + K_3 r^{\vartheta}) \qquad (r > 1)$$

(where $K_3 = K_1 \Sigma |a_n|^{-\vartheta}$). Hence

$$\log M(r) = O(r^h) + O(r^{\vartheta}) \text{ as } r \to \infty, \tag{20}$$

so that $\omega < \text{Max}(h, \vartheta)$. Since ϑ may be taken as near to τ as we please it follows that $\omega < \text{Max}(h, \tau)$.

If $\tau < h$, we may take $\vartheta < h$, and we conclude from (20) that (17) holds for $\beta = h = \text{Max}(h, \tau)$. If $\Sigma |a_n|^{-\tau}$ is a convergent infinite series, we may take $\vartheta = \tau$, and (17) holds for $\beta = \text{Max}(h, \vartheta) = \text{Max}(h, \tau)$.

Theorem F 2. $\tau \leqslant \omega$.

We may suppose that $G(z)$ has an infinity of zeros, since otherwise $\tau = 0 \leqslant \omega$. Apply Theorem D to $G(z)$ with $z_0 = 0$, $R = 2r_n$, $r = r_n$, where $r_n = |a_n|$. Since the n zeros $a_1, a_2, \ldots, a_n$ lie in $|z| \leqslant r_n$, we obtain

$$2^n \leqslant \frac{M(2r_n)}{|G(0)|}.$$

It follows, by the definition of ω, that, for any fixed positive ϵ,

$$n \log 2 = O\{(2r_n)^{\omega+\epsilon}\}$$

when $n \to \infty$. Hence, for all sufficiently large n,

$$n < r_n^{\,\omega+2\epsilon},$$

or

$$r_n^{\,-\omega-3\epsilon} < n^{-\frac{\omega+3\epsilon}{\omega+2\epsilon}}.$$

Since $(\omega + 3\epsilon)/(\omega + 2\epsilon) > 1$, it follows that $\Sigma r_n^{\,-\omega-3\epsilon}$ is convergent. Hence $\tau \leqslant \omega + 3\epsilon$, and therefore $\tau \leqslant \omega$, since ϵ may be taken arbitrarily small.

Theorem F 3. $h \leqslant \omega$.

The existence of ω implies, by Theorem F 2, that of τ, and therefore of Weierstrass's product (16). We have to show that $H(z)$ is a polynomial of degree not exceeding $[\omega]$. We may suppose $G(0) = 1$, $H(0) = 0$, since h and ω are not affected if we multiply $G(z)$ by a constant (different from 0). Let

$$H(z) = \sum_{1}^{\infty} b_\nu z^\nu.$$

Take a positive number R. Then

$$G(z) = \prod_{|a_n| < R} \left(1 - \frac{z}{a_n}\right) . G_R(z), \tag{21}$$

where $G_R(z)$ is an integral function of z having no zeros in $|z| < R$. Since $G_R(0) = G(0) = 1$ we can write

$$G_R(z) = e^{g_R(z)}, \tag{22}$$

where $g_R(z)$ is regular in $|z| < R$, and $g_R(0) = 0$. Let

$$g_R(z) = \sum_1^{\infty} c_\nu^{(R)} z^\nu \qquad (|z| < R).$$

On the circle $|z| = 2R$ we have

$$|G(z)| > \prod_{|a_n|<R} \left(\frac{2R}{R} - 1\right) . |G_R(z)| = |G_R(z)|,$$

and therefore $\qquad |G_R(z)| < M(2R).$

Since $G_R(z)$ is an integral function this holds, by the maximum modulus principle, for $|z| < 2R$, and in particular for $|z| < R$. It follows that

$$\Re g_R(z) = \log |G_R(z)| < \log M(2R) \qquad (|z| < R).$$

Hence by Theorem E (since $g_R(0) = 0$),

$$(23) \qquad |c_\nu^{(R)}| < \frac{2 \log M(2R)}{R^\nu} \qquad (\nu = 1, 2, \ldots).$$

Now comparing (21) and (22) with Weierstrass's product (16), we see that, for $|z| < R$,

$$H(z) + P_R(z) + \sum_{|a_n| \geq R} \left\{ \log\left(1 - \frac{z}{a_n}\right) + \frac{z}{a_n} + \ldots + \frac{1}{k}\left(\frac{z}{a_n}\right)^k \right\} = g_R(z),$$

where the logarithms have their principal values, and $P_R(z)$ is a polynomial of degree not higher than k. Differentiating ν times, dividing by $\nu!$, and putting $z = 0$, we deduce

$$(24) \qquad b_\nu - \frac{1}{\nu} \sum_{|a_n| \geq R} \frac{1}{a_n{}^\nu} = c_\nu^{(R)} \qquad (\nu > k).$$

Now suppose $\nu > \omega$. Then, by Theorem F 2, $\nu > \tau \geq k$. Hence by (23) and (24)

$$|b_\nu| < \frac{2 \log M(2R)}{R^\nu} + \frac{1}{\nu} \sum_{|a_n| \geq R} \frac{1}{|a_n|^\nu}.$$

In this inequality let $R \to \infty$, keeping ν fixed. The first term on the right tends to 0 because $\nu > \omega$, and the second because (ν being greater than τ) $\Sigma |a_n|^{-\nu}$ is a convergent (or finite) series. Since b_ν is independent of R, it follows that $b_\nu = 0$; and this is true for every $\nu > \omega$. Hence

$$H(z) = \sum_{\nu=1}^{[\omega]} b_\nu z^\nu,$$

which proves the theorem.

Theorem F 1 is due to Borel[1], Theorems F 2 and F 3 to Hadamard[2]. The above proof of Theorem F 3, which is much simpler than Hadamard's original proof, is due to Landau.[3] Hadamard's theory includes another set of relations involving the coefficients in the Taylor expansion of $G(z)$, but these do not concern us.

Combining Theorems F 1, F 2, F 3, we obtain the following result, which sums up the theory of integral functions so far as we shall require it.

Theorem F. *We have*

$$\omega = \text{Max}(\tau, h),$$

the existence of either side implying that of the other.

Further, if $\omega > 0$ *and if the relation* $\log M(r) = O(r^{\omega})$ *does not hold, then* $\tau = \omega$, $G(z)$ *has an infinity of zeros, and* $\Sigma |a_n|^{-\tau}$ *is divergent.*

The first part follows at once from the inequalities of Theorems F 1, F 2, F 3. To prove the second part we observe first that we must have $\tau = \omega$, since $\tau < \omega\ (= h)$ would imply $\log M(r) = O(r^{\omega})$ by F 1 (i). Next, since $\tau = \omega > 0$, $G(z)$ must have an infinity of zeros. Finally, convergence of the infinite series $\Sigma |a_n|^{-\tau}$ would imply $\log M(r) = O(r^{\omega})$ by F 1 (ii).

The numbers τ and ω are sometimes called respectively the 'real order' and the 'apparent order' of $G(z)$. The integer Max (k, h) is called the 'genus' of $G(z)$.

8. The zeros of $\xi(s)$. We shall first investigate the behaviour of the integral function $\xi(s)$ for large $|s|$, and determine its order ω.

Theorem 17. *If* $M(r)$ *is the maximum of* $|\xi(s)|$ *on the circle* $|s| = r$, *then*

$$\log M(r) \sim \tfrac{1}{2} r \log r$$

as $r \to \infty$.

We have

$$\xi(s) = \tfrac{1}{2} s (s-1) \pi^{-\frac{1}{2}s} \Gamma(\tfrac{1}{2}s) \zeta(s).$$

Now $|\zeta(s)| < \zeta(2)$ for $\sigma > 2$, and (by Theorem 9) $|\zeta(s)| < A_1 |t|^{\frac{1}{2}}$ for $\sigma > \frac{1}{2}$, $|t| > 2$; so that $|\zeta(s)| < A_2 |s|^{\frac{1}{2}}$ for $\sigma > \frac{1}{2}$, $|s| > 3$.

[1] E. Borel, *Leçons sur les fonctions entières* (Paris, Gauthier-Villars, 1900), 61.
[2] Hadamard 1.
[3] Landau 9; V, ii, 72–74.

Applying Stirling's Theorem[1] to $\Gamma(\frac{1}{2}s)$ we deduce that, for $\sigma \geqslant \frac{1}{2}$, $|s| = r > 3$,

$$|\xi(s)| < e^{|\frac{1}{2}s \log \frac{1}{2}s| + A_3|s|} < e^{\frac{1}{2}r\log r + A_4 r},$$

since $\log \frac{1}{2}s = \log |s| - \log 2 + i \arg s$ and $|\arg s| < \frac{1}{2}\pi$. From the functional equation $\xi(s) = \xi(1-s)$ we infer that

$$|\xi(s)| < e^{\frac{1}{2}|1-s|\log|1-s| + A_4|1-s|} < e^{\frac{1}{2}r\log r + A_5 r}$$

for $\sigma < \frac{1}{2}$, $|s| = r > 4$. Combining these inequalities we obtain

$$M(r) < e^{\frac{1}{2}r\log r + A_6 r} \qquad (r > 4).$$

On the other hand, if $r > 2$,

$$M(r) \geqslant \xi(r) > 1 \,.\, \pi^{-\frac{1}{2}r}\,\Gamma(\tfrac{1}{2}r) \,.\, 1 > e^{\frac{1}{2}r\log r - A_7 r}.$$

Thus we have, for $r > 4$,

$$\tfrac{1}{2}r\log r - A_7 r < \log M(r) < \tfrac{1}{2}r\log r + A_6 r,$$

whence the theorem.

Theorem 18. *$\xi(s)$ has an infinity of zeros. If these are denoted generally by ρ, the series $\sum_\rho |\rho|^{-\alpha}$ is convergent if $\alpha > 1$, divergent if $\alpha \leqslant 1$. Weierstrass's product for $\xi(s)$ is of the form*

$$\xi(s) = e^{b_0 + b_1 s} \prod_\rho \left\{\left(1 - \frac{s}{\rho}\right) e^{\frac{s}{\rho}}\right\},$$

where b_0 and b_1 are constants.

[1] A useful form to remember is

$$\log \Gamma(z+\alpha) = (z + \alpha - \tfrac{1}{2})\log z - z + \tfrac{1}{2}\log 2\pi + O(|z|^{-1}),$$

uniformly in any fixed angle $|\arg z| \leqslant \pi - \delta < \pi$ and any bounded range of α, as $|z| \to \infty$; the appropriate branches of the logarithms are, of course, those which are real for real positive values of the variable. This is an easy deduction from the special case $\alpha = 0$, but it is convenient for applications to retain the parameter α. For the proof (with $\alpha = 0$) see, e.g., Bieberbach's *Lehrbuch der Funktionentheorie* i, XIV.

We shall also require (though not in its complete form) the formula

$$\frac{\Gamma'(z)}{\Gamma(z)} = \log z - \frac{1}{2z} + O\left(\frac{1}{|z|^2}\right),$$

uniformly in $|\arg z| \leqslant \pi - \delta$; this may be proved in a similar way (but rather more simply) or deduced from the formula for $\log \Gamma(z)$ by means of the equation

$$f'(z) = \frac{1}{2\pi i}\int_C \frac{f(\zeta)}{(\zeta - z)^2}\,d\zeta,$$

where $f(z) = \log \Gamma(z) - (z - \frac{1}{2})\log z + z - \frac{1}{2}\log 2\pi$ and C is a circle with centre at $\zeta = z$ and radius $|z| \sin \frac{1}{2}\delta$.

Also

$$(25)\qquad \frac{\xi'}{\xi}(s) = b_1 + \sum_\rho \left(\frac{1}{s-\rho} + \frac{1}{\rho}\right),$$

$$(26)\qquad \frac{\zeta'}{\zeta}(s) = b - \frac{1}{s-1} - \tfrac{1}{2}\frac{\Gamma'}{\Gamma}(\tfrac{1}{2}s+1) + \sum_\rho \left(\frac{1}{s-\rho} + \frac{1}{\rho}\right),$$

where $b = b_1 + \frac{1}{2}\log \pi$.[1]

Since $\xi(0) = \frac{1}{2} \neq 0$ (Theorem 15), the theory of § 7 is immediately applicable to $\xi(s)$. By Theorem 17, $\xi(s)$ is of order 1, and the relation $\log M(r) = O(r)$ does not hold. Hence, by Theorem F, $\tau = 1$, $h = 1$ or 0, $\xi(s)$ has an infinity of zeros, and $\Sigma|\rho|^{-1}$ summed over these zeros is divergent. This proves the assertions about $\Sigma|\rho|^{-\alpha}$ and Weierstrass's product. Equation (25) is obtained by logarithmic differentiation of the product, and (26) follows in virtue of the relation

$$\xi(s) = (s-1)\pi^{-\frac{1}{2}s}\Gamma(\tfrac{1}{2}s+1)\zeta(s).$$

It is, of course, a consequence of Weierstrass's theory that the infinite product and series are absolutely convergent for all s (distinct from the ρ's in the case of the series).

It can be shown that

(27) $b_1 = \frac{1}{2}\log(4\pi) - 1 - \frac{1}{2}C = -0{\cdot}023\ldots$, $b = \log(2\pi) - 1 - \frac{1}{2}C = 0{\cdot}549\ldots$,

where C is Euler's constant[2]; but the actual values are not required for our main applications.

9. We denote a typical zero of $\xi(s)$, that is to say a typical non-trivial zero of $\zeta(s)$, by

$$\rho = \beta + \gamma i.$$

We know from Theorem 16 that (for each ρ) $0 \leqslant \beta \leqslant 1$, and from Theorem 10 that $\beta < 1$, so that, by the symmetry about $\sigma = \frac{1}{2}$, $0 < \beta < 1$. We shall now carry these results a stage further by proving

Theorem 19. *There exists a positive absolute constant a such that $\zeta(s)$ has no zeros in the domain*[3]

$$\sigma > 1 - \frac{a}{\log(|t|+2)}.$$

[1] We have adopted a compact notation for logarithmic derivatives.

[2] See T, 3–4.

[3] It is, of course, the order of magnitude of $a/\log(|t|+2)$ for large $|t|$ that is important; but we write $\log(|t|+2)$ instead of $\log|t|$ in order to obtain a formula valid for all t. This device is constantly employed in this kind of work.

The proof is based, like that of Theorem 10 (which we do not need to assume), on the inequality

$$(28) \qquad 3+4\cos\theta+\cos 2\theta \geqslant 0.$$

We start from (26), which we write as

$$(29) \qquad f(s)=g(s)+\frac{\zeta'}{\zeta}(s),$$

where

$$f(s)=\sum_\rho\left(\frac{1}{s-\rho}+\frac{1}{\rho}\right), \quad g(s)=-b+\frac{1}{s-1}+\tfrac{1}{2}\frac{\Gamma'}{\Gamma}(\tfrac{1}{2}s+1).$$

Using the formula $\zeta'(s)/\zeta(s)=-\Sigma\Lambda(n)n^{-s}$ in conjunction with (28), we obtain, for $\sigma>1$ and real t,

$$\Re\left(3\frac{\zeta'}{\zeta}(\sigma)+4\frac{\zeta'}{\zeta}(\sigma+ti)+\frac{\zeta'}{\zeta}(\sigma+2ti)\right)\leqslant 0.$$

Hence, by (29),

$$(30) \quad \Re\{3f(\sigma)+4f(\sigma+ti)+f(\sigma+2ti)\}$$
$$\leqslant\Re\{3g(\sigma)+4g(\sigma+ti)+g(\sigma+2ti)\} \quad (\sigma>1).$$

Now choose (as is clearly possible) a positive number a_1 so that $\zeta(s)$ has no zeros in the square $|\sigma-1|\leqslant a_1$, $|t|\leqslant a_1$, and suppose in what follows that $1<\sigma\leqslant 2$, $|t|>a_1$. Then, using the asymptotic formula for Γ'/Γ (p. 57, footnote), we obtain

$$(31) \quad \Re\{3g(\sigma)+4g(\sigma+ti)+g(\sigma+2ti)\}<\frac{3}{\sigma-1}+A_1\log(|t|+2).$$

On the other hand we have, for $s=\sigma,\ \sigma+ti,\ \sigma+2ti$,

$$\Re f(s)=\sum_\rho\Re\left(\frac{1}{s-\rho}+\frac{1}{\rho}\right)=\sum_\rho\left(\frac{\sigma-\beta}{|s-\rho|^2}+\frac{\beta}{|\rho|^2}\right)>\frac{\sigma-\beta_0}{|s-\rho_0|^2}>0,$$

where $\rho_0=\beta_0+\gamma_0 i$ is any selected zero; for $\sigma>1$, and $0\leqslant\beta\leqslant 1$ for each ρ. Hence

$$(32) \quad \Re\{3f(\sigma)+4f(\sigma+ti)+f(\sigma+2ti)\}>4\frac{(\sigma-\beta_0)}{(\sigma-\beta_0)^2+(t-\gamma_0)^2}.$$

By (32), (30), (31),

$$4\frac{(\sigma-\beta_0)}{(\sigma-\beta_0)^2+(t-\gamma_0)^2}<\frac{3}{\sigma-1}+A_1\log(|t|+2)$$

for $1<\sigma\leqslant 2$, $|t|>a_1$; and the restriction $\sigma\leqslant 2$ may now be dropped (subject possibly to an increase in the value of A_1), since the left-hand side is not greater than $4/(\sigma-\beta_0)<4$ if $\sigma>2$. We deduce in particular, putting $t=\gamma_0$ and dropping the suffix 0,

$$\frac{4}{\sigma-\beta}-\frac{3}{\sigma-1}<A_2\log(|\gamma|+2) \qquad (\sigma>1), \tag{33}$$

where $\rho=\beta+\gamma i$ is any zero with $|\gamma|>a_1$.

From (33) we deduce first that $\beta<1$; for if $\beta=1$ the left-hand side tends to infinity when $\sigma\to 1+0$. Hence, for any $\sigma>1$, we can write $\sigma=1+\lambda(1-\beta)$, where $\lambda>0$; (33) then becomes

$$\left(\frac{4}{1+\lambda}-\frac{3}{\lambda}\right)\frac{1}{1-\beta}<A_2\log(|\gamma|+2) \qquad (\lambda>0).$$

Since $4>3$, the expression in brackets on the left is positive when λ is large enough. Taking, for example, $\lambda=4$, we obtain

$$\beta<1-\frac{a_2}{\log(|\gamma|+2)},$$

where $a_2=1/20A_2$. This being true for every ρ with $|\gamma|>a_1$, the theorem follows, in virtue of our choice of a_1; we may take, for example, $a=\text{Min}(a_2, a_1\log 2)$.

10. In making deductions from Theorem 19 we shall replace the special function $a/\log(|t|+2)$ by a general function satisfying stated conditions. We shall then be in a position to infer at once the consequences of possible refinements of Theorem 19. We first require a theorem on the order of magnitude of ζ'/ζ.

Theorem 20. *Suppose that $\zeta(s)$ has no zeros in the domain*

$$\sigma>1-\eta(|t|),$$

where $\eta(t)$ is, for $t\geqslant 0$, a decreasing function, having a continuous derivative $\eta'(t)$, and satisfying the following conditions:

(i) $0<\eta(t)\leqslant\frac{1}{2}$,

(ii) $\eta'(t)\to 0$ *as* $t\to\infty$,

(iii) $\dfrac{1}{\eta(t)}=O(\log t)$ *as* $t\to\infty$.

Let α be a fixed number satisfying $0<\alpha<1$.

Then

$$\frac{\zeta'}{\zeta}(s) = O(\log^2 |t|),$$

uniformly in the region

$$\sigma \geqslant 1 - \alpha\eta(|t|),$$

when $t \to \pm\infty$.

We may suppose $t > 0$. And we may confine ourselves to the range $1 - \alpha\eta(t) \leqslant \sigma \leqslant 1 + \alpha\eta(t)$ of σ; for, if $\sigma \geqslant 1 + \alpha\eta(t)$, we have, writing $\eta = \eta(t)$,

$$\left|\frac{\zeta'}{\zeta}(s)\right| < \Sigma \frac{\Lambda(n)}{n^{1+\alpha\eta}} = -\frac{\zeta'}{\zeta}(1+\alpha\eta) < \frac{A_1}{\alpha\eta},$$

since ζ'/ζ has a simple pole at the point 1; and this is $O(\log t)$ by (iii).

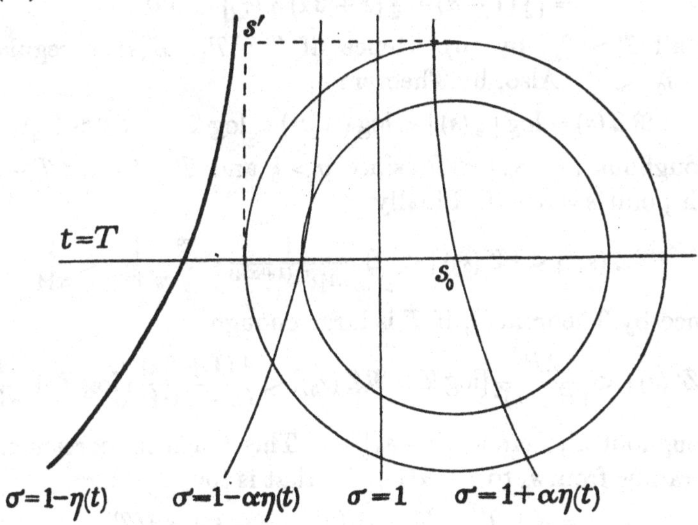

Fig. 4

Since $\zeta(s)$ is regular and not zero in the simply-connected domain D defined by $t > 0$, $\sigma > 1 - \eta(t)$, there is a branch $Z(s)$ of $\log\zeta(s)$ regular in D and defined by

$$Z(s) = \log\zeta(s) = \sum_{p,m} \frac{1}{mp^{ms}}$$

(p. 17, (13)) when $\sigma > 1$. We apply Theorem E to $Z(s)$ and to two circles (see Fig. 4) having a common centre at $s_0 = 1 + \alpha H + Ti$

and passing through the points $1-\alpha\mathrm{H}+Ti$ and $1-\frac{1}{2}(1+\alpha)\mathrm{H}+Ti$ respectively, where $T>1$ and $\mathrm{H}=\eta(T)$; the radii of these circles are

$$r=2\alpha\mathrm{H}, \qquad R=\alpha\mathrm{H}+\tfrac{1}{2}(1+\alpha)\mathrm{H}=\tfrac{1}{2}(1+3\alpha)\mathrm{H}.$$

When T is large enough these circles lie in D. For, since $R<2\mathrm{H}\leqslant 1$ (by (i)), $T>1$, and $\eta(t)$ is a decreasing function, this will certainly be so if the point

$$1-\tfrac{1}{2}(1+\alpha)\mathrm{H}+(T+R)i=\sigma'+t'i=s'$$

lies in D; and this condition, which is equivalent to $\sigma'>1-\eta(t')$, is fulfilled for large T, since

$$\begin{aligned}\sigma'-1+\eta(t')&=-\tfrac{1}{2}(1+\alpha)\mathrm{H}+\eta(T+R)\\ &=-\tfrac{1}{2}(1+\alpha)\mathrm{H}+\eta(T)+R\eta'(\tau) \qquad (T<\tau<T+R)\\ &=\{\tfrac{1}{2}(1-\alpha)+\tfrac{1}{2}(1+3\alpha)\eta'(\tau)\}\mathrm{H}>0\end{aligned}$$

for all $T>T_1$, by (ii). Hence, if $T>T_1$, $Z(s)$ is regular in $|s-s_0|<R$. Also, by Theorem 9,

$$\Re Z(s)=\log|\zeta(s)|<\log(A_2t^{\frac{1}{2}})<\log T \qquad (T>T_2)$$

throughout $|s-s_0|<R$, since $\sigma>\frac{1}{2}$ and $T-1<t<T+1$ at each point $s=\sigma+ti$. Finally

$$|\Re Z(s_0)|\leqslant|Z(s_0)|\leqslant\sum_{p,m}\frac{1}{mp^{m(1+\alpha\mathrm{H})}}<\sum_{2}^{\infty}\frac{1}{n^{1+\alpha\mathrm{H}}}<\frac{1}{\alpha\mathrm{H}}.$$

Hence by Theorem E, if T is large enough,

$$|Z'(s)|<\frac{2R}{(R-r)^2}\{\log T-\Re Z(s_0)\}<\frac{4(1+3\alpha)}{(1-\alpha)^2\mathrm{H}}\left(\log T+\frac{1}{\alpha\mathrm{H}}\right)$$

throughout the circle $|s-s_0|\leqslant r$. This holds in particular on the radius from s_0 to $1-\alpha\mathrm{H}+Ti$, that is for

$$s=\sigma+Ti, \qquad 1-\alpha\eta(T)\leqslant\sigma\leqslant 1+\alpha\eta(T).$$

Since $Z'=\zeta'/\zeta$, and $1/\mathrm{H}=O(\log T)$ by (iii), the theorem follows.

11. Application to $\psi(x)$ and $\pi(x)$. We can now obtain improved forms of the asymptotic relations for $\pi(x)$ and the associated functions. As in Chapter II we begin with $\psi_1(x)$.

Theorem 21. *Under the conditions of Theorem 20, we have*

$$\psi_1(x)=\tfrac{1}{2}x^2+O(x^2e^{-\alpha\omega(x)})$$

as $x\to\infty$, where $\omega(x)$ is the minimum of $\eta(t)\log x+\log t$ for $t\geqslant 1$.

The minimum $\omega(x)$ exists for every $x > 0$, since $\eta(t)\log x + \log t$ is continuous for $t \geq 1$ and tends to infinity with t.

Suppose $x > 1$. Starting from the fundamental formula (14), p. 31, and applying Cauchy's theorem, we obtain

$$\frac{\psi_1(x)}{x^2} = \tfrac{1}{2} - \frac{1}{2\pi i}\int_C \frac{x^{s-1}}{s(s+1)}\frac{\zeta'}{\zeta}(s)\,ds = \tfrac{1}{2} + J,$$

say, where C is the curve $\sigma = 1 - \alpha\eta(|t|)$; for in the region bounded by C and the line $\sigma = c$ the integrand is regular, except at $s = 1$, and is, by Theorem 20, uniformly

$$O(t^{-2}\log^2|t|) = o(1),$$

when $t \to \pm\infty$ (x fixed). In estimating J we may by symmetry confine ourselves to C_1, the upper half of C. Using Theorem 20, and denoting by $K_1, K_2, \ldots$ positive numbers depending only on the function η and the number α, we have, on C_1,

$$\left|\frac{\zeta'}{\zeta}(s)\right| < K_1\log^2(t+2), \qquad \left|\frac{ds}{dt}\right| = |-\alpha\eta'(t) + i| < K_2,$$

whence

$$|J| < K_3\int_0^\infty \frac{x^{-\alpha\eta(t)}\log^2(t+2)}{(t+1)^2}\,dt = K_3\int_1^\infty \frac{x^{-\alpha\eta(u-1)}\log^2(u+1)}{u^2}\,du.$$

Hence, since $\eta(u-1) \geq \eta(u)$ and $x > 1$,

$$\left|\frac{\psi_1(x)}{x^2} - \frac{1}{2}\right| = |J| < K_3\int_1^\infty e^{-\alpha\eta(u)\log x - \alpha\log u}\frac{\log^2(u+1)}{u^{2-\alpha}}\,du$$

$$< K_3 e^{-\alpha\omega(x)}\int_1^\infty \frac{\log^2(u+1)}{u^{2-\alpha}}\,du = K_4 e^{-\alpha\omega(x)},$$

the last integral being convergent since $2 - \alpha > 1$.

Theorem 22. *With the conditions and notation of Theorems* 20 *and* 21, *we have*

$$\psi(x) = x + O(xe^{-\frac{1}{2}\alpha\omega(x)}), \tag{34}$$

$$\pi(x) = \operatorname{li} x + O(xe^{-\frac{1}{2}\alpha\omega(x)}). \tag{35}$$

We begin by noting some simple properties of $\omega(x)$. Let $x_1 > x_2 > 0$ and let t_1 and t_2 be values of t for which the minima $\omega(x_1)$ and $\omega(x_2)$ are attained. Then

$$\omega(x_2) \leq \eta(t_1)\log x_2 + \log t_1 < \eta(t_1)\log x_1 + \log t_1 = \omega(x_1),$$

$$\omega(x_1) \leq \eta(t_2)\log x_1 + \log t_2$$
$$= \omega(x_2) + \eta(t_2)(\log x_1 - \log x_2) < \omega(x_2) + (\log x_1 - \log x_2),$$

so that $\omega(x)$ and $\log x - \omega(x)$ are (strictly) increasing functions of x for $x > 0$. Since $\omega(1) = 0$ and $\log 1 - \omega(1) = 0$, we infer in particular that $0 < \omega(x) < \log x$ for $x > 1$.

Now suppose $x > 2$. Let h be a function of x satisfying $0 < h < \frac{1}{2}x$. Since ψ is an increasing function,

$$(36) \qquad \frac{1}{h}\int_{x-h}^{x} \psi(u)\,du < \psi(x) < \frac{1}{h}\int_{x}^{x+h} \psi(u)\,du.$$

The outside expressions are

$$(37) \qquad \frac{\psi_1(x \mp h) - \psi_1(x)}{\mp h} = x \mp \frac{h}{2} + O\left(\frac{x^2 e^{-\alpha\omega(\frac{1}{2}x)}}{h}\right),$$

by Theorem 21, since $\frac{1}{2}x < x - h < x < x + h < \frac{3}{2}x$ and ω is an increasing function. In the O-term we may replace $\omega(\frac{1}{2}x)$ by $\omega(x)$; for, since $\log u - \omega(u)$ is an increasing function of u,

$$e^{-\alpha\omega(\frac{1}{2}x)} = (\tfrac{1}{2}x)^{-\alpha} e^{\alpha\{\log\frac{1}{2}x - \omega(\frac{1}{2}x)\}} < (\tfrac{1}{2}x)^{-\alpha} e^{\alpha\{\log x - \omega(x)\}} = 2^{\alpha} e^{-\alpha\omega(x)}.$$

Hence, by (36) and (37),

$$\psi(x) = x + O(h) + O(h^{-1}x^2 e^{-\alpha\omega(x)}).$$

Taking $h = \frac{1}{2}xe^{-\frac{1}{2}\alpha\omega(x)}$, we obtain (34).

We next deduce by partial summation the corresponding formula for $\Pi(x)$ (p. 18, (16)). We have first

$$\Pi(x) = \sum_{2 \leq n \leq x} \frac{\Lambda(n)}{\log n} = \int_2^x \frac{\psi(u)\,du}{u \log^2 u} + \frac{\psi(x)}{\log x}$$

by Theorem A (p. 18). Also

$$\int_2^x \frac{du}{\log u} = \int_2^x \frac{u\,du}{u\log^2 u} + \frac{x}{\log x} - \frac{2}{\log 2}$$

by partial integration. Hence, by subtraction,

$$(38) \quad \Pi(x) - \operatorname{li} x = \int_2^x \frac{\psi(u) - u}{u \log^2 u}\,du + \frac{\psi(x) - x}{\log x} + O(1).$$

Using (34) we deduce

$$\Pi(x) - \operatorname{li} x = O\left(\int_2^x e^{-\frac{1}{2}\alpha\omega(u)}\,du\right) + O(xe^{-\frac{1}{2}\alpha\omega(x)}) + O(1).$$

The term $O(1)$ on the right may be omitted since

$$(39) \qquad xe^{-\frac{1}{2}\alpha\omega(x)} > xe^{-\frac{1}{2}\alpha\log x} = x^{1-\frac{1}{2}\alpha} > x^{\frac{1}{2}} > 1.$$

And the integral is

$$\int_2^x e^{\frac{1}{2}\alpha\{\log u-\omega(u)\}}\frac{du}{u^{\frac{1}{2}\alpha}} < e^{\frac{1}{2}\alpha\{\log x-\omega(x)\}}\int_2^x \frac{du}{u^{\frac{1}{2}\alpha}} < \frac{xe^{-\frac{1}{2}\alpha\omega(x)}}{1-\frac{1}{2}\alpha},$$

since $1-\frac{1}{2}\alpha > 0$. Hence

$$\Pi(x) - \operatorname{li} x = O(xe^{-\frac{1}{2}\alpha\omega(x)}). \tag{40}$$

Finally

$$\Pi(x) - \pi(x) = \sum_{m=2}^{M} \frac{\pi(x^{1/m})}{m} \qquad \left(M = \left[\frac{\log x}{\log 2}\right]\right)$$

$$= O(x^{\frac{1}{2}}) + O(Mx^{\frac{1}{3}}) = O(x^{\frac{1}{2}}), \tag{41}$$

and (35) follows from (40), (41), and (39).

12. We now specialise the results of § 11 by means of Theorem 19. Theorem 22 gives

Theorem 23. *When* $x \to \infty$,

$$\psi(x) = x + O(xe^{-a\sqrt{(\log x)}}), \tag{42}$$

$$\pi(x) = \operatorname{li} x + O(xe^{-a\sqrt{(\log x)}}), \tag{43}$$

where a is a positive absolute constant.

For by Theorem 19 we can choose the $\eta(t)$ of §§ 10 and 11 so that $\eta(t) = a_1/(\log t)$ for $t > 2$, if a_1 is suitably chosen. Then, if $x > 1$, we have, using the fact that $u + v \geqslant 2\sqrt{(uv)}$ for $u, v > 0$,

$$\eta(t)\log x + \log t \geqslant \begin{cases} 2\sqrt{(a_1 \log x)} & (t \geqslant 2) \\ \eta(2)\log x & (1 < t < 2), \end{cases}$$

the first relation being an equality if $\log t = \sqrt{(a_1 \log x)}$. Hence $\omega(x) = 2\sqrt{(a_1 \log x)}$ for all sufficiently large x. Taking $\alpha = \frac{1}{2}$ in Theorem 22 we deduce Theorem 23 with $a = \frac{1}{2}\sqrt{a_1}$.

The function $e^{a\sqrt{(\log x)}}$ increases more rapidly than any positive power of $\log x$ but less rapidly than any positive power of x. Thus (43) implies that

$$\pi(x) = \operatorname{li} x + O\left(\frac{x}{\log^{\Delta} x}\right) \tag{44}$$

for any fixed positive Δ, but does not allow us to infer that the error term is $O(x^{1-\delta})$, where $\delta > 0$.

It is easily proved, by partial integration, that

$$\operatorname{li} x = \frac{x}{\log x} + \frac{1!\,x}{\log^2 x} + \dots + \frac{(k-1)!\,x}{\log^k x} + O\left(\frac{x}{\log^{k+1} x}\right),$$

where k is any fixed positive integer. Using this in (44), and taking $\Delta = 4$, $k = 3$, we obtain, after a little reduction,

$$\pi(x) - \frac{x}{\log x - B} = (1 - B)\frac{x}{\log^2 x} + (2 - B^2)\frac{x}{\log^3 x} + O\left(\frac{x}{\log^4 x}\right),$$

where B is any constant. From these results it is clear that Legendre's function $x/(\log x - B)$ will give the best approximation to $\pi(x)$ for large x if we take $B = 1$, and that even then this function is not so good an approximation as $\operatorname{li} x$. (Cf. Chapter I, § 8, p. 21.)

13. The results of §§ 9 and 12 are due to de la Vallée Poussin; the above method of deducing Theorem 23 from Theorem 19 is due to Landau.[1] Landau has given a proof of Theorem 19 (and so of Theorem 23) not depending on the theory of integral functions, or indeed on the existence of $\zeta(s)$ in the whole plane.[2]

De la Vallée Poussin gave a numerical value for the constant a in (43), and this was improved by Landau. But there is no object in reproducing these values, since it is now known that the function $\sqrt{(\log x)}$ is itself capable of improvement. This advance is due to Littlewood. He has gone beyond Theorem 19 and shown that $\zeta(s)$ has no zeros in a domain of the form

$$\sigma > 1 - a_1 \frac{\log\log(|t| + 3)}{\log(|t| + 3)}.$$

This theorem lies very deep, and for the proof we must refer to Titchmarsh's tract.[3] But we can easily obtain the resulting refinement of Theorem 23. For we can choose the $\eta(t)$ of §§ 10 and 11 so that

$$\eta(t) = a_2 \frac{\log\log t}{\log t} \quad \text{for } t \geqslant 3.$$

Then, if $\log\xi = \sqrt{(\log x)}$ and x is so large that $\xi \geqslant 3$, we have

$$\eta(t)\log x + \log t \geqslant \begin{cases} 2(a_2 \log x \log\log t)^{\frac{1}{2}} \geqslant (2a_2 \log x \log\log x)^{\frac{1}{2}} & (t \geqslant \xi) \\ \eta(\xi)\log x = \frac{1}{2}a_2 (\log x)^{\frac{1}{2}} \log\log x & (1 \leqslant t \leqslant \xi). \end{cases}$$

Hence $\omega(x) \geqslant (2a_2 \log x \log\log x)^{\frac{1}{2}}$ for all sufficiently large x. Thus Theorem 22 gives

Theorem 24. *When* $x \to \infty$,

$$\psi(x) = x + O\left(xe^{-a\sqrt{\log x \log\log x}}\right),$$

$$\pi(x) = \operatorname{li} x + O\left(xe^{-a\sqrt{\log x \log\log x}}\right),$$

where a is a positive absolute constant.

These are the best known results of their kind.[4]

[1] De la Vallée Poussin 2; H, i, 318–333.

[2] V, ii, 9–28; T, 14–17 (Theorem 8).

[3] T, 20–23 (Theorem 13); V, ii, 31–44. For the original statement see Littlewood 2 (§ 4). Landau's method shows to such advantage here that Littlewood's proof has never been published.

[4] BC, 789, footnote [176)]; V, ii, 3–8, 44–47.

14. The Riemann hypothesis. Riemann conjectured that the non-trivial zeros of $\zeta(s)$ all lie on the line $\sigma = \frac{1}{2}$, but this has never been proved or disproved. The evidence for the truth of the 'Riemann hypothesis' (as this assertion is now called), and its relation to the general theory of $\zeta(s)$, are discussed in the companion tract on the zeta-function, to which we refer the reader for further information.[1] It falls within the scope of this tract, however, to examine the bearing of the hypothesis on the problem of the distribution of primes. We can see at once that the asymptotic relations of the present chapter admit of substantial improvement if the hypothesis is true. For in this case we can take the $\eta(t)$ of §§ 10 and 11 to be $\frac{1}{2}$, and, taking $\alpha = 1 - \epsilon$ $(0 < \epsilon < 1)$, we obtain errors $O(x^{\frac{3}{2}+\frac{1}{2}\epsilon})$ in the formula for $\psi_1(x)$ and $O(x^{\frac{3}{4}+\frac{1}{4}\epsilon})$ in the formulae for $\psi(x)$ and $\pi(x)$. But these results can be improved still further. In the first place the index which we should expect in the latter formulae, corresponding to the $\frac{3}{2}$ in the former, is not $\frac{3}{4}$ but $\frac{1}{2}$, and the increase results merely from the imperfections of the method; moreover, the factors $x^{\frac{1}{2}\epsilon}$ and $x^{\frac{1}{4}\epsilon}$ can be replaced by powers of $\log x$. While these defects can, to some extent, be removed by appropriate modification of the arguments of this chapter, it is preferable to use a different method. This is based on the 'explicit formulae' which form the main subject of the next chapter.

[1] T, Chapters III and V. See also Siegel **1**.

CHAPTER IV

EXPLICIT FORMULAE

1. In this chapter we propose to discuss a curious type of exact representation, by infinite series, of functions associated with $\psi(x)$. These 'explicit formulae' are very remarkable and interesting in themselves, and have important applications, some of which will be dealt with at the end of the chapter.

Our arguments will involve applications of Cauchy's theorem in which a line of integration is moved across the critical strip, and we must first obtain more precise information about the distribution of the imaginary parts of the complex zeros of $\zeta(s)$.

2. Density of zeros. We denote by $N(T)$, where $T > 0$, the number (necessarily finite) of zeros of $\zeta(s)$ in the rectangle $0 < \sigma < 1$, $0 < t < T$, that is, by Theorem 16, the number of zeros $\rho = \beta + \gamma i$ of $\zeta(s)$, or of $\xi(s)$, for which $0 < \gamma < T$.

Theorem 25. *When* $T \to \infty$,

$$N(T) = \frac{T}{2\pi} \log \frac{T}{2\pi} - \frac{T}{2\pi} + O(\log T).$$

Suppose that $T > 3$, and (for the present) that T is not equal to any γ. Then $\xi(s)$ has $2N(T)$ zeros inside the rectangle whose vertices are $2 \pm Ti$ and $-1 \pm Ti$, and none on its boundary. Hence, by a theorem of Cauchy (the 'principle of the argument'),

$$4\pi N(T) = [\arg \xi(s)]_C,$$

where $[\arg \xi(s)]_C$ denotes the increase in $\arg \xi(s)$ when s describes the perimeter C of this rectangle in the positive sense. Now

$$[\arg \xi(s)]_C = [\arg \tfrac{1}{2} s(s-1)]_C + [\arg \phi(s)]_C,$$

where $\phi(s) = \pi^{-\frac{1}{2}s} \Gamma(\tfrac{1}{2}s) \zeta(s)$. The first term on the right is 4π, and, since $\phi(s)$ takes equal values at points s and $1 - s$ and conjugate values at points $\sigma \pm ti$, the second term is clearly $4[\arg \phi(s)]_L$, where L is the broken line made up of the segment L_1 from 2 to $2 + Ti$ followed by the segment L_2 from $2 + Ti$ to $\tfrac{1}{2} + Ti$. Hence

$$(1) \quad \pi N(T) = \pi + [\arg \pi^{-\frac{1}{2}s}]_L + [\arg \Gamma(\tfrac{1}{2}s)]_L + [\arg \zeta(s)]_L.$$

We have first

$$[\arg \pi^{-\frac{1}{2}s}]_L = [-\tfrac{1}{2}t\log\pi]_L = -\tfrac{1}{2}T\log\pi.$$

Next (see p. 57, footnote, with $z = \frac{1}{2}Ti$, $\alpha = \frac{1}{4}$)

$$\begin{aligned}[\arg\Gamma(\tfrac{1}{2}s)]_L &= [\mathfrak{I}\log\Gamma(\tfrac{1}{2}s)]_L = \mathfrak{I}\log\Gamma(\tfrac{1}{4}+\tfrac{1}{2}Ti) - \mathfrak{I}\log\Gamma(1)\\ &= \mathfrak{I}\{(-\tfrac{1}{4}+\tfrac{1}{2}Ti)\log(\tfrac{1}{2}Ti) - \tfrac{1}{2}Ti + \tfrac{1}{2}\log 2\pi\} + O(T^{-1})\\ &= \tfrac{1}{2}T\log(\tfrac{1}{2}T) - \tfrac{1}{8}\pi - \tfrac{1}{2}T + O(T^{-1})\end{aligned}$$

when $T\to\infty$. Substituting into (1), we obtain

$$(2)\quad N(T) = \frac{T}{2\pi}\log\frac{T}{2\pi} - \frac{T}{2\pi} + \frac{7}{8} + \frac{1}{\pi}[\arg\zeta(s)]_L + O\left(\frac{1}{T}\right).$$

Now let m be the number (necessarily finite, as will appear) of distinct points s' of L (excluding end-points) at which $\mathfrak{R}\zeta(s) = 0$. Then

$$(3)\qquad [\arg\zeta(s)]_L \leqslant (m+1)\pi;$$

for, when s describes one of the $m+1$ pieces into which L is divided by the points s', $\arg\zeta(s)$ cannot vary by more than π since $\mathfrak{R}\zeta(s)$ does not change sign. Now no point s' can lie on L_1, since

$$(4)\qquad \mathfrak{R}\zeta(2+ti) \geqslant 1 - \sum_2^\infty \frac{1}{n^2} > 1 - \frac{1}{2^2} - \int_2^\infty \frac{du}{u^2} = \frac{1}{4}.$$

Thus m is the number of distinct points σ of the interval $\frac{1}{2} < \sigma < 2$ at which $\mathfrak{R}\zeta(\sigma+Ti) = 0$, and this is the number of distinct zeros of the function $g(s) = \frac{1}{2}\{\zeta(s+Ti) + \zeta(s-Ti)\}$ on the segment $\frac{1}{2} < s < 2$ of the real axis; for $g(\sigma) = \mathfrak{R}\zeta(\sigma+Ti)$ for real σ, since $\zeta(\sigma \pm Ti)$ are conjugate. Since $g(s)$ is regular, except at $s = 1 \pm Ti$, m is therefore finite, and we obtain an upper bound for m by applying Theorem D (p. 49) to $g(s)$ and to the circles $|s-2| \leqslant \frac{7}{4}$, $|s-2| \leqslant \frac{3}{2}$. Since $T > 3$, $g(s)$ is regular in the larger circle and satisfies

$$|g(s)| < \tfrac{1}{2}A_1(|t+T|^{\frac{1}{4}} + |t-T|^{\frac{1}{4}}) < A_1(T+2)^{\frac{1}{4}}$$

by Theorem 9, since $\sigma \geqslant \frac{1}{4}$ and $1 < |t \pm T| < 2 + T$ at all points $s = \sigma + ti$ of the circle. Also $g(2) = \mathfrak{R}\zeta(2+Ti) > \frac{1}{4}$ by (4). Hence, by Theorem D,

$$\left(\frac{7}{6}\right)^m < \frac{A_1(T+2)^{\frac{1}{4}}}{\frac{1}{4}} < T \qquad (T > T_0 \geqslant 3).$$

Thus $m < A_2 \log T$ for $T > T_0$. Substituting into (2) and (3) we deduce

$$\left| N(T) - \frac{T}{2\pi}\log\frac{T}{2\pi} + \frac{T}{2\pi} \right| < A_3 \log T$$

for $T > T_0$, provided that T is not equal to a γ. But this restriction is now irrelevant, as may be seen by replacing T by a larger value T' (distinct from the γ's) and making $T' \to T + 0$.

We note some consequences of Theorem 25 which will be useful in applications.

Theorem 25 *a*. *If h is a fixed positive number, then*

$$N(T+h) - N(T) = O(\log T)$$

when $T \to \infty$.

For, writing

$$P(t) = (t/2\pi)\log(t/2\pi) - (t/2\pi),$$

we have

$$P(T+h) - P(T) = hP'(T + \vartheta h) \quad (0 < \vartheta < 1),$$

and the result follows from Theorem 25, since

$$P'(t) = (1/2\pi)\log(t/2\pi).$$

Theorem 25 *b*. *We have*

$$\sum_{0<\gamma\leqslant T} \frac{1}{\gamma} = O(\log^2 T), \quad \sum_{\gamma > T} \frac{1}{\gamma^2} = O\left(\frac{\log T}{T}\right)$$

as $T \to \infty$.

The summations extend over all ρ whose imaginary parts γ lie within the specified ranges, due allowance being made for multiplicity. Denoting the sums by S and S' respectively, we have

$$S < \sum_{m=0}^{[T]} s_m, \quad S' < \sum_{m=[T]}^{\infty} s_m',$$

where s_m and s_m' are $\Sigma 1/\gamma$ and $\Sigma 1/\gamma^2$ summed over the range $m < \gamma \leqslant m+1$. If $m \geqslant 1$, the number of terms in s_m or s_m' is

$$N(m+1) - N(m) = \nu_m,$$

say, and $s_m \leqslant \nu_m/m$, $s_m' \leqslant \nu_m/m^2$. By Theorem 25*a*, $\nu_m = O(\log m)$ as $m \to \infty$; hence, when $T \to \infty$,

$$S = O(1) + O\left(\sum_{2}^{[T]} \frac{\log m}{m}\right) = O(\log^2 T),$$

$$S' = O\left(\sum_{[T]}^{\infty} \frac{\log m}{m^2}\right) = O\left(\frac{\log T}{T}\right).$$

Theorem 25 c. *If the* $\rho = \beta + \gamma i$ *with* $\gamma > 0$ *are arranged in a sequence* $\rho_n = \beta_n + \gamma_n i$ *so that* $\gamma_{n+1} \geqslant \gamma_n$, *then*

$$|\rho_n| \sim \gamma_n \sim \frac{2\pi n}{\log n} \qquad \text{as } n \to \infty.$$

For, since $N(\gamma_n - 1) < n \leqslant N(\gamma_n + 1)$ and

$$2\pi N(\gamma_n \pm 1) \sim (\gamma_n \pm 1) \log(\gamma_n \pm 1) \sim \gamma_n \log \gamma_n,$$

we have first $2\pi n \sim \gamma_n \log \gamma_n$, whence $\log n \sim \log \gamma_n$, and

$$\gamma_n \sim 2\pi n / \log \gamma_n \sim 2\pi n / \log n.$$

And $\gamma_n \leqslant |\rho_n| < \gamma_n + 1$.

Theorem 25, which embodies one of the most important properties of $\zeta(s)$, was stated by Riemann, but first proved by von Mangoldt.[1] The difficult step is, of course, the estimation of $[\arg \zeta(s)]_L$. The above proof is due to Backlund.[2] Unlike von Mangoldt's proof it does not depend on Theorem 18, and it thus provides an alternative proof of the existence of an infinity of complex zeros. The information obtained in this way about the density of distribution of the zeros is more precise than that contained in Theorem 18; for we now see (from Theorem 25 c) that $\Sigma |\rho|^{-1} (\log |\rho|)^{-\alpha}$ is convergent for $\alpha > 2$, divergent for $\alpha \leqslant 2$.

It should be observed that Theorem 25 *a*, which suffices for many applications, is essentially simpler than Theorem 25. It may be proved at once by applying Theorem D directly to $\zeta(s)$ and to two circles with centre at $c + Ti$ and passing through the points $\frac{1}{2} + (T + 2h)i$ and $\frac{1}{2} + (T + h)i$ respectively, where $c = c(h)$ is a sufficiently large positive number, and using the symmetry about $\sigma = \frac{1}{2}$.

3. We next obtain an inequality for $|\zeta'/\zeta|$ on a certain set of lines crossing the critical strip and avoiding the zeros of $\zeta(s)$.

Theorem 26. *There exists a sequence of numbers* $T_2, T_3, \ldots$, *such that*

$$m < T_m < m + 1 \qquad (m = 2, 3, \ldots)$$

and

$$\left| \frac{\zeta'}{\zeta}(s) \right| < A \log^2 t \qquad (-1 \leqslant \sigma \leqslant 2, t = T_m).$$

We have by Theorem 18, (26),

$$\frac{\zeta'}{\zeta}(s) = b - \frac{1}{s-1} - \frac{1}{2} \frac{\Gamma'}{\Gamma}(\tfrac{1}{2}s + 1) + \sum_\rho \left(\frac{1}{s - \rho} + \frac{1}{\rho} \right) = g(s) + \Sigma(s),$$

say, where $\Sigma(s)$ stands for the infinite series. Let $s = \sigma + ti$,

[1] Riemann 1; von Mangoldt 1, 2, 4; H, i, 368–378.

[2] Backlund 1, 2. For further developments see T, 58–61, 87–93, 96.

$s_0 = 2 + ti$, where $-1 < \sigma < 2$, $t > 2$, and t is not equal to a γ. Let δ_0 be the distance of t from the nearest γ, and let

$$\delta = \delta(t) = \operatorname{Min}(\delta_0, 1).$$

Then for every $\rho = \beta + \gamma i$

$$|s - \rho|^2 > (t-\gamma)^2 > \tfrac{1}{2}\delta^2 + \tfrac{1}{2}(t-\gamma)^2 > \tfrac{1}{2}\delta^2\{1 + (t-\gamma)^2\},$$
$$|s_0 - \rho|^2 = (2-\beta)^2 + (t-\gamma)^2 > 1 + (t-\gamma)^2,$$

since $0 < \beta < 1$. Hence

$$|\Sigma(s) - \Sigma(s_0)| = \left|\sum_\rho \frac{s_0 - s}{(s-\rho)(s_0-\rho)}\right|$$
$$< \sum_\rho \frac{3}{(\frac{1}{2}\delta^2)^{\frac{1}{2}}\{1+(t-\gamma)^2\}} < \frac{6}{\delta}\sum_\rho \frac{1}{1+(t-\gamma)^2}.$$

On the other hand

$$\Re\Sigma(s_0) = \sum_\rho\left(\frac{2-\beta}{|s_0-\rho|^2} + \frac{\beta}{|\rho|^2}\right) > \sum_\rho \frac{1}{4+(t-\gamma)^2} > \tfrac{1}{4}\sum_\rho \frac{1}{1+(t-\gamma)^2},$$

since $0 < \beta < 1$, $|s_0 - \rho|^2 < 4 + (t-\gamma)^2$. Hence

$$|\Sigma(s) - \Sigma(s_0)| < \frac{24}{\delta}\Re\Sigma(s_0) < \frac{24}{\delta}|\Sigma(s_0)|.$$

Thus

$$\left|\frac{\zeta'}{\zeta}(s) - g(s)\right| = |\Sigma(s)| < \frac{25}{\delta}|\Sigma(s_0)| = \frac{25}{\delta}\left|\frac{\zeta'}{\zeta}(s_0) - g(s_0)\right|.$$

Applying the asymptotic formula for Γ'/Γ (p. 57, footnote) to $g(s)$ and $g(s_0)$, and noting that $|\zeta'(s_0)/\zeta(s_0)| < \Sigma\Lambda(n)n^{-2}$, we deduce

$$\left|\frac{\zeta'}{\zeta}(s)\right| < \frac{A_1}{\delta}\log t. \tag{5}$$

Now let m be any integer greater than 1, and ν_m the number of ρ for which $m < \gamma < m+1$, so that $\nu_m = N(m+1) - N(m)$. Then, if we divide the interval $(m, m+1)$ into $\nu_m + 1$ equal parts, one subinterval at least will contain no γ in its interior; take T_m to be the mid-point of such an interval. Then, if $t = T_m$, we have $\delta > \{2(\nu_m + 1)\}^{-1}$, and the theorem follows from (5), since (by Theorem 25 *a*) $\nu_m < A_2 \log m < A_2 \log t$.

We shall require also the following theorem, which is more elementary.

Theorem 27. *In the region obtained by removing from the half-plane* $\sigma < -1$ *the interiors of a set of circles of radius* $\frac{1}{2}$ *with centres at* $s = -2, -4, -6, \ldots$, *i.e. in the region defined by*

$$\sigma < -1, \ |s - n| > \tfrac{1}{2} \qquad (n = -2, -4, -6, \ldots),$$

we have

$$\left|\frac{\zeta'}{\zeta}(s)\right| < A \log(|s| + 1).$$

Changing s to $1 - s$ in the functional equation and differentiating logarithmically, we obtain

$$\frac{\zeta'}{\zeta}(s) = \log 2\pi + \tfrac{1}{2}\pi \cot \tfrac{1}{2}\pi s - \frac{\Gamma'}{\Gamma}(1 - s) - \frac{\zeta'}{\zeta}(1 - s).$$

Now, when s is in the region R defined above, we have $1 - \sigma > 2$, so that

$$\left|\frac{\zeta'}{\zeta}(1 - s)\right| < \Sigma \frac{\Lambda(n)}{n^2},$$

$$\left|\frac{\Gamma'}{\Gamma}(1 - s)\right| < A_1 |\log(1 - s)| < A_2 \log(|s| + 1).$$

Further (see p. 43, footnote)

$$|\tfrac{1}{2}\pi \cot \tfrac{1}{2}\pi s| = \left|\frac{\pi i}{2} + \frac{\pi i}{e^{\pi s i} - 1}\right| < A_3$$

in R. Combining these inequalities we obtain the desired result.

4. Explicit formula for $\psi_1(x)$. We proceed to the discussion of explicit formulae, and we begin as usual with $\psi_1(x)$.

Theorem 28. *If* $x > 1$, *then*

$$\psi_1(x) = \frac{x^2}{2} - \sum_\rho \frac{x^{\rho+1}}{\rho(\rho + 1)} - x \frac{\zeta'}{\zeta}(0) + \frac{\zeta'}{\zeta}(-1) - \sum_{r=1}^{\infty} \frac{x^{1-2r}}{2r(2r - 1)}.$$

Consider the integral

$$I(m) = -\frac{1}{2\pi i} \int_C \frac{x^{s+1}}{s(s + 1)} \frac{\zeta'}{\zeta}(s)\, ds$$

taken in the positive sense round the rectangle $C = C(m)$ whose vertices are $2 \pm T_m i$, $-2m - 1 \pm T_m i$, where m is an integer greater than 1 and the T_m are the numbers of Theorem 26. Let $I_1(m)$ be the integral along the side $(2 - T_m i, 2 + T_m i)$ and $I_2(m)$ the integral along the remainder of C.

Now, when $m \to \infty$, $I_1(m) \to \psi_1(x)$ by the fundamental formula (14), p. 31. In $I_2(m)$ we have, by Theorems 26 and 27,

$$\left|\frac{\zeta'}{\zeta}(s)\right| < A_1 \log^2 |s| < A_2 \log^2 m,$$

since $|s| < 2m + 1 + T_m < 3m + 2$; also $|s| > m$, $|s+1| > m$, and $|x^{s+1}| = x^{\sigma+1} < x^3$ since $x \geqslant 1$. Hence

$$|I_2(m)| < \frac{x^3 . A_2 \log^2 m}{2\pi m^2}(4m + 6 + 2T_m) < A_3 x^3 \frac{\log^2 m}{m},$$

so that $I_2(m) \to 0$ as $m \to \infty$. It follows that

$$\text{(6)} \qquad I(m) \to \psi_1(x) \quad \text{as} \quad m \to \infty.$$

But by Cauchy's theorem of residues

$$I(m) = \frac{x^2}{2} - \sum_{|\gamma| < T_m} \frac{x^{\rho+1}}{\rho(\rho+1)} - x\frac{\zeta'}{\zeta}(0) + \frac{\zeta'}{\zeta}(-1) - \sum_{r=1}^{m} \frac{x^{-2r+1}}{(-2r)(-2r+1)};$$

for the poles of the integrand are at the points 1, 0, -1 and the zeros (ρ and $-2r$) of $\zeta(s)$, and a zero of order n of $\zeta(s)$ gives rise to a simple pole with residue n of $\zeta'(s)/\zeta(s)$, and a pole of order n to a simple pole with residue $-n$. Making $m \to \infty$ and comparing with (6) we obtain the theorem. The two infinite series are absolutely convergent, since the general terms are less in modulus than x^2/γ^2 and $1/r^2$ respectively. The second series may, of course, be summed in finite terms, but this is unimportant.

A generalised form of Theorem 28 was the basis of de la Vallée Poussin's proof of the prime number theorem.[1] He proved the formula by substituting for $-\zeta'(s)/\zeta(s)$ from equation (26), p. 58, and integrating term-by-term—a procedure not difficult to justify. But this method is not available for the more delicate problem to be discussed in the following sections.

5. The explicit formula for $\psi_1(x)$ is based ultimately on the special case $k = 1$ of Theorem B (p. 31). The general form of this theorem leads to an explicit formula for

$$\psi_k(x) = \frac{1}{k!}\sum_{n \leqslant x}(x-n)^k \Lambda(n) = \frac{1}{(k-1)!}\int_0^x (x-u)^{k-1}\psi(u)\,du,$$

[1] De la Vallée Poussin 1.

the kth repeated integral (from 0 to x) of $\psi(x)$. The function $\psi(x)$ itself may be regarded as corresponding formally to $k = 0$, but this case presents theoretical difficulties, because the denominator $s(s+1)\dots(s+k)$, which secures absolute convergence of the integral in Theorem B when $k > 0$, reduces to the single factor s when $k = 0$. We must first prove an analogue of Theorem B for $k = 0$.

Theorem G. *If $c > 0$, $y > 0$, then*

$$\frac{1}{2\pi i}\int_{c-\infty i}^{c+\infty i}\frac{y^s}{s}\,ds = \begin{cases} 0 & (y<1), \\ \frac{1}{2} & (y=1), \\ 1 & (y>1), \end{cases} \tag{7}$$

where, in the case $y = 1$, the integral is to be interpreted as a 'Cauchy principal value', that is to say as the limit of

$$\frac{1}{2\pi i}\int_{c-Ti}^{c+Ti}\frac{y^s}{s}\,ds, \tag{8}$$

when $T \to \infty$.

Moreover, if $I(y) = I(y, T) + \Delta(y, T)$,

where $I(y)$ and $I(y, T)$ denote the integrals (7) *and* (8) *respectively, then, for $T > 0$,*

$$|\Delta(y,T)| < \begin{cases} \dfrac{y^c}{\pi T\,|\log y|} & (y \neq 1), \\ \dfrac{c}{\pi T} & (y=1), \end{cases} \tag{9}$$

$$|\Delta(y,T)| < y^c \qquad (\textit{always}). \tag{10}$$

Suppose first $y > 1$. By Cauchy's theorem

$$\frac{1}{2\pi i}\int_C \frac{y^s}{s}\,ds = 1,$$

the integral being taken in the positive sense round the rectangle whose vertices are $c - Ui$, $c + Vi$, $-X + Vi$, $-X - Ui$, where $U > 0$, $V > 0$, $X > 0$. Let $X \to \infty$, keeping U and V fixed. Then the integral along the side $(-X + Vi, -X - Ui)$ tends to 0; for the path is of fixed length $U + V$ and $|y^s/s| < y^{-X}/X < 1/X$ on the path, since $y > 1$, $X > 0$. Hence

$$\frac{1}{2\pi i}\int_{c-Ui}^{c+Vi}\frac{y^s}{s}\,ds = 1 - \frac{1}{2\pi i}\int_{-\infty-Ui}^{c-Ui}\frac{y^s}{s}\,ds + \frac{1}{2\pi i}\int_{-\infty+Vi}^{c+Vi}\frac{y^s}{s}\,ds$$
$$= 1 - J(-U) + J(V),$$

say; $J(-U)$ and $J(V)$ are absolutely convergent, and

$$|J(-U)| < \frac{1}{2\pi}\int_{-\infty}^{c} \frac{y^\sigma}{U}\,d\sigma = \frac{y^c}{2\pi U \log y}, \quad |J(V)| < \frac{y^c}{2\pi V \log y},$$

since $y^\sigma = e^{\sigma \log y}$ and $\log y > 0$. These results establish both (7) and (9) in this case. The case $y < 1$ is treated similarly, except that we use a rectangle lying to the right of the line $\sigma = c$.

For the case $y = 1$ we have

$$\frac{1}{2\pi i}\int_{c-Ti}^{c+Ti} \frac{ds}{s} = \frac{1}{2\pi}\int_{-T}^{T} \frac{dt}{c+ti} = \frac{1}{2\pi}\int_{-T}^{T} \frac{c-ti}{c^2+t^2}\,dt = \frac{1}{\pi}\int_{0}^{T} \frac{c\,dt}{c^2+t^2},$$

since the real and imaginary parts of the integrand are respectively an even and an odd function of t. Hence

$$\left|\frac{1}{2\pi i}\int_{c-Ti}^{c+Ti} \frac{ds}{s} - \frac{1}{2}\right| = \frac{1}{\pi}\int_{T}^{\infty} \frac{c\,dt}{c^2+t^2} < \begin{cases} c/\pi T, \\ \tfrac{1}{2}, \end{cases}$$

the first inequality being obtained by replacing the denominator $c^2 + t^2$ by t^2, and the second by replacing the lower limit T by 0. This establishes (7), (9), (10), when $y = 1$.

It remains to prove (10) when $y \neq 1$. Let Γ be the arc of the circle $|s| = (c^2 + T^2)^{\frac{1}{2}} = R$ lying to the left or right of the line $\sigma = c$ according as $y > 1$ or $y < 1$. Then, by Cauchy's theorem and (7),

$$|\Delta(y, T)| = \left|\frac{1}{2\pi i}\int_{\Gamma} \frac{y^s}{s}\,ds\right| < \frac{1}{2\pi}\cdot\frac{y^c}{R}\cdot 2\pi R = y^c.$$

It is easy to show that, for $U > 0$, $V > 0$,

$$\frac{1}{2\pi i}\int_{c-Ui}^{c+Vi} \frac{ds}{s} = \frac{1}{2} - \frac{i}{2\pi}\log\frac{V}{U} + \vartheta\frac{2c}{\pi T} \qquad (|\vartheta| < 1), \tag{11}$$

where $T = \mathrm{Min}\,(U, V)$, so that the integral (7) is not convergent in the ordinary sense when $y = 1$.

6. Explicit formula for $\psi_0(x)$. Theorem G will be found to lead to an explicit formula, not for $\psi(x)$ itself, but for the function

$$\psi_0(x) = \frac{\psi(x+0) + \psi(x-0)}{2} = \sum_{n \leqslant x}{}' \Lambda(n),$$

where Σ' indicates that, when x is an integer, the term corresponding to $n = x$ is to have the factor $\frac{1}{2}$; $\psi_0(x)$ differs from $\psi(x)$ only when x is a prime power p^m, the difference then being $\frac{1}{2}\log p$.

Theorem 29. *If $x > 1$, then*

$$(12) \qquad \psi_0(x) = x - \sum_\rho \frac{x^\rho}{\rho} - \frac{\zeta'}{\zeta}(0) - \tfrac{1}{2}\log\left(1 - \frac{1}{x^2}\right),$$

where $\sum_\rho$ means the limit of

$$(13) \qquad S(x, T) = \sum_{|\gamma| \leqslant T} \frac{x^\rho}{\rho}$$

when $T \to \infty$.

Further, if

$$(14) \qquad \sum_\rho \frac{x^\rho}{\rho} = S(x, T) + R(x, T),$$

then, for $x > 1$, $T > 3$,

$$(15) \quad |R(x, T)| < \begin{cases} A \dfrac{x^3}{x-1}\left(\dfrac{\log^2 T}{T} + \dfrac{1}{T\xi}\right) & (x \neq p^m), \\[2ex] A \dfrac{x^3}{x-1} \dfrac{\log^2 T}{T} & (x = p^m), \end{cases}$$

$$(16) \quad |R(x, T)| < A \frac{x^3}{x-1} \frac{\log^2 T}{T} + A \log x \qquad (\textit{always}),$$

where $\xi = \xi(x)$ is the distance of x from the nearest prime power p^m.

Suppose $T > 3$ and let T' be the T_m of Theorem 26 next greater than T. Let q be an odd integer greater than T, and consider

$$J(q) = \frac{1}{2\pi i} \int_C \frac{x^s}{s}\left(-\frac{\zeta'}{\zeta}(s)\right) ds = \frac{1}{2\pi i} \int_C f(s)\, ds$$
$$= J_1 + J_2(q) + J_3(q) + J_4(q),$$

say, where the integral is taken in the positive sense round the rectangle with vertices $2 \pm T'i$, $-q \pm T'i$, and J_1, J_2, J_3, J_4 are the contributions of the four sides taken in order starting with the side $(2 - T'i, 2 + T'i)$. In J_1 we substitute from the equation $-\zeta'(s)/\zeta(s) = \Sigma\Lambda(n)n^{-s}$ and integrate term-by-term (as is clearly permissible by uniform convergence); this gives, in the notation of Theorem G, with $c = 2$,

$$J_1 = \sum_1^\infty \Lambda(n) I\left(\frac{x}{n}, T'\right) = \sum_1^\infty \Lambda(n) I\left(\frac{x}{n}\right) - \sum_1^\infty \Lambda(n)\Delta\left(\frac{x}{n}, T'\right)$$
$$= \psi_0(x) - \sum_1^\infty \Lambda(n)\Delta\left(\frac{x}{n}, T'\right) = \psi_0(x) - X,$$

say, by (7) and the definition of $\psi_0(x)$. On the three remaining sides of C we have, by Theorems 26 and 27,

$$|f(s)| < \frac{x^\sigma}{|s|} A_1 \log^2 |s| < 3A_1 \frac{\log^2 3|s|}{3|s|} x^\sigma$$

$$(17) \qquad < 3A_1 \frac{\log^2 3T}{3T} x^\sigma < A_2 \frac{\log^2 T}{T} x^\sigma,$$

since $3|s| > 3T > 9 > e^2$ and $(\log^2 u)/u$ is a decreasing function for $u > e^2$. Hence $|J_3(q)| < K_T x^{-q}$ (where K_T depends only on T), so that, since $x > 1$, $J_3(q) \to 0$ as $q \to \infty$. Thus

$$\lim_{q\to\infty} J(q) = \psi_0(x) - X - \frac{1}{2\pi i}\int_{-\infty}^{2} \{f(\sigma + T'i) - f(\sigma - T'i)\} d\sigma$$

$$(18) \qquad = \psi_0(x) - X - Y,$$

say. The integral Y is absolutely convergent by (17) (since $x > 1$), and

$$(19) \quad |Y| < \frac{A_2}{\pi} \frac{\log^2 T}{T} \int_{-\infty}^{2} e^{\sigma \log x} d\sigma = A_3 \frac{\log^2 T}{T} \frac{x^2}{\log x}.$$

On the other hand by the theory of residues

$$\lim_{q\to\infty} J(q) = x - \sum_{|\gamma| < T'} \frac{x^\rho}{\rho} - \frac{\zeta'}{\zeta}(0) - \sum_{r=1}^{\infty} \frac{x^{-2r}}{-2r}.$$

Comparing with (18) and summing the infinite series, we obtain, using the notation (13),

$$\psi_0(x) = x - S(x, T') - \frac{\zeta'}{\zeta}(0) - \tfrac{1}{2}\log\left(1 - \frac{1}{x^2}\right) + X + Y$$

$$(20) \qquad = x - S(x, T) - \frac{\zeta'}{\zeta}(0) - \tfrac{1}{2}\log\left(1 - \frac{1}{x^2}\right) + P,$$

where

$$(21) \quad P = X + Y + \{S(x, T) - S(x, T')\} = X + Y + Z,$$

say. Since $T < T' < T + 2$, we have

$$(22) \qquad |Z| < 2\{N(T') - N(T)\} \frac{x}{T} < A_4 \frac{\log T}{T} x$$

by Theorem 25 *a*.

We have now to estimate

$$(23) \qquad X = \sum_1^\infty \Lambda(n) \Delta\left(\frac{x}{n}, T'\right) = \sum_1^\infty u_n,$$

say. By Theorem G, (9), since $c=2$,

$$(24)\quad |u_n| < \Lambda(n)\frac{(x/n)^2}{\pi T'\,|\log(x/n)|} < \frac{x^2}{\pi T}\frac{\Lambda(n)}{n^2}\frac{n+x}{|n-x|}\qquad (n \neq x);$$

for, if $v = \mathrm{Min}\,(x/n, n/x)$, then $0 < v < 1$ and

$$\left|\log\frac{x}{n}\right| = -\log\{1-(1-v)\} > 1-v > \frac{1-v}{1+v} = \frac{|n-x|}{n+x}.$$

Let $\nu = \nu(x)$ be the integer defined by $\nu - \frac{1}{2} < x < \nu + \frac{1}{2}$. If $|n-x| > \frac{1}{2}x$, then

$$\frac{n+x}{|n-x|} = \left|1 + \frac{2x}{n-x}\right| \leqslant 1 + \frac{2x}{|n-x|} < 5.$$

If $|n-x| \leqslant \frac{1}{2}x$ but $n \neq \nu$, we can write $n = \nu \pm r$, where r is an integer satisfying

$$0 < r = |n-\nu| \leqslant |n-x| + |\nu - x| \leqslant 2\,|n-x| \leqslant x.$$

Hence, by (23) and (24),

$$|X - u_\nu| \leqslant \frac{x^2}{\pi T}\left(\sum_{|n-x|>\frac{1}{2}x}\frac{\log n}{n^2}\,5 + 2\sum_{r=1}^{[x]}\frac{\log(\frac{3}{2}x)}{(\frac{1}{2}x)^2}\frac{\frac{5}{2}x}{\frac{1}{2}r}\right)$$

$$(25)\qquad < \frac{x^2}{\pi T}\left(5\sum_{1}^{\infty}\frac{\log n}{n^2} + A_5\frac{\log^2(2x)}{x}\right) < A_6\frac{x^2}{T}.$$

By (21), (25), (19), (22),

$$|P - u_\nu| < A_6\frac{x^2}{T} + A_3\frac{\log^2 T}{T}\frac{x^2}{\log x} + A_4\frac{\log T}{T}x$$

$$(26)\qquad < A_7\frac{\log^2 T}{T}\left(\frac{x^2}{\log x} + x^2\right).$$

It remains to consider u_ν. Suppose first that x is not a p^m. Then if ν is a p^m, we have by (24), since $\nu + x \leqslant 2\nu + \frac{1}{2} < 3\nu$,

$$|u_\nu| < \frac{x^2}{\pi T}\frac{3\Lambda(\nu)}{\nu}\frac{1}{|\nu - x|} < A_8\frac{x^2}{T\xi},$$

since $|\nu - x| = \xi$ by the definition of ξ; and the inequality is true also if ν is not a p^m since $u_\nu = 0$ in this case. If x is a p^m, then

$$|u_\nu| = |u_x| = |\Lambda(x)\Delta(1, T')| < \log x\,.\,\frac{2}{\pi T'} < A_9\frac{x^2}{T}$$

by Theorem G, (9). And in any case, by Theorem G, (10),

$$|u_\nu| \leqslant (\log\nu)\,(x/\nu)^2 < A_{10}\log x.$$

Combining these results with (26), and noting that

$$\log x = -\log\{1-(1-x^{-1})\} > 1-x^{-1}, \qquad 1 > 1-x^{-1},$$

we see that P satisfies inequalities similar to (15) and (16). This proves the theorem. For the inequalities corresponding to (15), taken in conjunction with (20), show first of all that, for any fixed $x>1$, $S(x,T)$ tends to a limit $S(x)$ as $T\to\infty$, and that $S(x)=S(x,T)-P$; and we then conclude from (14) that $R(x,T)=-P$, so that the desired inequalities for $R(x,T)$ follow from those established for P.

The inequalities (15) and (16) are of course susceptible of improvement. The order in x can evidently be reduced by replacing the line $\sigma=2$ in the proof by a line lying further to the left. And there is an alternative treatment of the horizontal parts of the contour which reduces the order in T by a factor $\log T$.[1] But the forms given are sufficient for our applications.

7. The 'explicit formula' (12) suggests that there are connections between the numbers p^m (the discontinuities of $\psi_0(x)$) and the numbers ρ. But no relationship essentially more explicit than (12) has ever been established between these two sets of numbers.

The inequalities for the 'remainder' $R(x,T)$ give interesting information about the behaviour of the series $\Sigma x^\rho/\rho$. Write, as in Theorem 25 *c*, $\rho_n=\beta_n+\gamma_n i$ $(n=1,2,\ldots;\ \gamma_{n+1}\geqslant\gamma_n>0)$, and let $\rho_{-n}=\beta_n-\gamma_n i$. Write $v_n(x)=x^{\rho_n}/\rho_n$ $(n=\pm1,\pm2,\ldots)$, and consider the series

$$(27)\qquad \sum_1^\infty\{v_n(x)+v_{-n}(x)\}=\sum_1^\infty 2\Re v_n(x).$$

If $S_N(x)$ is the sum of the first N terms, and $S(x)=\lim\limits_{T\to\infty}S(x,T)$, then

$$|S_N(x)-S(x)|\leqslant|S_N(x)-S(x,\gamma_N)|+|R(x,\gamma_N)|<\frac{2x\,\nu_N}{\gamma_N}+|R(x,\gamma_N)|,$$

where ν_N is the number of $\rho=\beta+\gamma i$ with $\gamma=\gamma_N$. Since $\nu_N=O(\log\gamma_N)$ by Theorem 25 *a*, it follows at once from (15) that (27) is convergent (to sum $S(x)$) for all $x>1$, and uniformly convergent in any closed interval lying within this range and containing no prime power. The convergence cannot be uniform in an interval containing a prime power, since $\psi_0(x)$, and therefore $S(x)$, is discontinuous at the points $x=p^m$. The series is, however, 'boundedly convergent' in any fixed interval $1<a\leqslant x\leqslant b$, as may be seen from (16). When $x\leqslant 1$, the reasoning of § 6 breaks down, but we can infer the convergence properties of (27) for

[1] V, ii, 108–120; Landau 4. For a deeper result, on the assumption of the Riemann hypothesis, see Littlewood 2 (§ 8), 3.

$0<x<1$ from those for $x>1$ by observing that $\rho'=1-\rho$ describes with ρ the non-trivial zeros of $\zeta(s)$ and that

$$\left|\frac{x^\rho}{\rho}+x\frac{x^{-\rho'}}{\rho'}\right|=\left|\frac{x^\rho}{\rho\rho'}\right|<\frac{1}{\gamma^2}\qquad(0<x<1);$$

the conclusion is that (27) is convergent for $0<x<1$, the points p^{-m} playing a similar rôle to the points p^m. The explicit formula (12) is, of course, no longer valid, but by treating the integral

$$\frac{1}{2\pi i}\int_C\frac{x^{1-s}}{s-1}\left(-\frac{\zeta'}{\zeta}(s)\right)ds$$

by the methods of § 6 we obtain the explicit formula

$$\sum_{n\leq 1/x}{}'\frac{\Lambda(n)}{n}=\log\frac{1}{x}-C+\sum_\rho\frac{x^\rho}{\rho}-x+\tfrac{1}{2}\log\frac{1+x}{1-x}\qquad(0<x<1),$$

where C is Euler's constant. The series (27) may be shown to exhibit a 'Gibbs phenomenon' in the neighbourhood of the points $p^{\pm m}$. For $x=1$, (27) is (absolutely) convergent, but neither explicit formula holds, and the series cannot be boundedly convergent near $x=1$ (on either side) owing to the presence of the terms $-\tfrac{1}{2}\log(1-x^{-2})$ and $-\tfrac{1}{2}\log\{(1+x)/(1-x)\}$ in the explicit formulae.

The behaviour of the separate series

$$\sum_1^\infty v_n(x),\quad \sum_1^\infty v_{-n}(x)$$

can be investigated by using in the argument of § 6 a rectangle

$$(2-U'i,\ 2+V'i,\ -q+V'i,\ -q-U'i)$$

with $U'\neq V'$, and applying the corresponding extension of Theorem G. It is found that the series are uniformly convergent in any closed interval lying in $x>0$ and containing none of the points 1, $p^{\pm m}$, but are not convergent at these points.[1]

We may note a close analogy with the familiar formula

$$[x]_0-x+\tfrac{1}{2}=\frac{1}{\pi}\sum_1^\infty\frac{\sin 2\pi nx}{n}=\sum_{-\infty}^{\infty}{}''\frac{e^{2\pi nxi}}{2\pi ni}\qquad\left(\Sigma''=\sum_{n\neq 0}\right),$$

which may be regarded as an explicit formula for the function

$$[x]_0=\tfrac{1}{2}([x+0]+[x-0])=\sum_{n\leq x}{}'1\qquad(x>0),$$

and may indeed be proved by similar (though naturally simpler) methods with the aid of the generating function

$$\frac{1}{e^s-1}=\sum_1^\infty e^{-ns}\qquad(\sigma>0).$$

Riemann's fundamental memoir centred round an explicit formula analogous to (12) for the function $\Pi_0(x)=\tfrac{1}{2}\{\Pi(x+0)+\Pi(x-0)\}$ (in our notation). His formula (after correction of a numerical error and with a change of notation) is

$$\Pi_0(x)=\mathrm{li}\,x-\sum_\rho\mathrm{li}\,x^\rho+\int_x^\infty\frac{du}{(u^2-1)u\log u}-\log 2\qquad(x>1),$$

[1] Landau 5; Cramér 1, 2; T, 61–63.

where $\mathrm{li}\, x^{\rho} = \mathrm{li}\, e^{\rho \log x}$ and

$$\mathrm{li}\, e^{w} = \int_{-\infty+vi}^{u+vi} \frac{e^z}{z}\, dz \qquad (w = u + vi,\ v \gtrless 0).[1]$$

This and the formula for $\psi_0(x)$ were first proved rigorously by von Mangoldt.[2]

8. Applications. As a first application we shall show how the explicit formula for $\psi_1(x)$ (Theorem 28) may be employed in the proof of the prime number theorem. We have

$$(28)\qquad \frac{\psi_1(x)}{x^2} = \tfrac{1}{2} - \sum_{\rho} \frac{x^{\rho-1}}{\rho(\rho+1)} + O\left(\frac{1}{x}\right) = \tfrac{1}{2} - U(x) + O\left(\frac{1}{x}\right),$$

say, when $x \to \infty$. Now the series $U(x)$ is uniformly convergent over the range $x \geq 1$, since

$$\left|\frac{x^{\rho-1}}{\rho(\rho+1)}\right| = \frac{x^{\beta-1}}{|\rho(\rho+1)|} < \frac{x^{\beta-1}}{\gamma^2} < \frac{1}{\gamma^2}$$

and $\Sigma \gamma^{-2}$ is convergent. Hence, when $x \to \infty$,

$$\lim U(x) = \sum_{\rho} \lim \frac{x^{\rho-1}}{\rho(\rho+1)} = \sum_{\rho} 0 = 0,$$

since $|x^{\rho-1}| = x^{\beta-1}$ and $\beta < 1$ in each term by Theorem 10. Thus $\psi_1(x)/x^2 \to \frac{1}{2}$, whence the prime number theorem follows by Theorem C (p. 35) as before.

The formula (28) may also be used in conjunction with Theorems 19 and $25a$ to prove the more precise relations of Theorem 23. This is, in essentials, the method adopted by de la Vallée Poussin.[3]

For our further applications we introduce a number Θ defined as follows:

Θ is the upper bound of the real parts of the zeros of $\zeta(s)$.

Clearly $\Theta \leq 1$, since there are no zeros in $\sigma > 1$. And from the existence of the non-trivial zeros ρ and their symmetry about the line $\sigma = \frac{1}{2}$ we infer that $\Theta \geq \frac{1}{2}$. Thus $\frac{1}{2} \leq \Theta \leq 1$, and this is the most that is known about Θ; but $\Theta = \frac{1}{2}$ if (and only if) the Riemann hypothesis is true. We now prove the following

[1] If $\mathrm{li}\, e^{u}$ is defined for real u as $\frac{1}{2} \lim_{v \to 0} (\mathrm{li}\, e^{u+vi} + \mathrm{li}\, e^{u-vi})$, it is easily verified that this agrees with the definition of $\mathrm{li}\, x$ given on p. 3. Our definition of $\mathrm{li}\, e^{w}$ makes this function continuous on the stretch $w < 0$ of the real axis, and differs from the current definition, which makes it continuous on the stretch $w > 0$. Our $\mathrm{li}\, e^{w}$ is asymptotically e^{w}/w when $w \to \infty$ in any fixed angle $\delta \leq \arg w \leq 2\pi - \delta$ $(0 < \delta < \pi)$.

[2] von Mangoldt 2; H, i, 333–368 (see also 516–532); Cramér 1.

[3] De la Vallée Poussin 2.

theorem, which is worthless if $\Theta = 1$, but supersedes the results of Chapter III if $\Theta < 1$.

Theorem 30. *We have*

$$\psi_1(x) = \tfrac{1}{2}x^2 + O(x^{\Theta+1}), \tag{29}$$

$$\psi(x) = x + O(x^{\Theta}\log^2 x), \tag{30}$$

$$\pi(x) = \operatorname{li} x + O(x^{\Theta}\log x). \tag{31}$$

The relation (29) is an immediate consequence of (28); for

$$|U(x)| < \sum_{\rho} \frac{x^{\beta-1}}{\gamma^2} < \sum_{\rho} \frac{x^{\Theta-1}}{\gamma^2} = A_1 x^{\Theta-1} \qquad (x > 1),$$

since $\Sigma\gamma^{-2}$ is convergent.

To prove (30) we use Theorem 29, taking $T = x^2$ $(x > 2)$ in (16). This gives

$$\psi_0(x) = x - S(x, x^2) - R(x, x^2) + O(1) = x - S(x, x^2) + O(\log^2 x).$$

Now
$$|S(x, x^2)| < 2 \sum_{0<\gamma\leqslant x^2} \frac{x^{\Theta}}{\gamma} = O(x^{\Theta}\log^2 x)$$

by Theorem 25 *b*. Since $\psi(x) - \psi_0(x) = O(\log x)$ the result follows.

Finally (31) is deduced by partial summation. Using (30) in conjunction with equations (38) and (41) on pp. 64–65, we obtain

$$\pi(x) - \operatorname{li} x = O\left(\int_2^x u^{\Theta-1}\,du\right) + O(x^{\Theta}\log x) + O(x^{\frac{1}{2}}) = O(x^{\Theta}\log x),$$

since $\Theta > \frac{1}{2}$.

There is an alternative method which leads to (30) in a less direct but more elementary way, since it depends on Theorem 28 instead of Theorem 29.[1] Let h be a function of x such that $1 < h < \frac{1}{2}x$ for large x. Then by Theorem 28

$$\frac{\psi_1(x \pm h) - \psi_1(x)}{\pm h} = x \pm \tfrac{1}{2}h - \sum_{\rho} \frac{(x \pm h)^{\rho+1} - x^{\rho+1}}{\rho(\rho+1)(\pm h)} - \frac{\zeta'}{\zeta}(0) + O\left(\frac{1}{xh}\right). \tag{32}$$

Denoting the general term of the series on the right by w_ρ we have

$$|w_\rho| < \frac{(x+h)^{\Theta+1} + x^{\Theta+1}}{\gamma^2 h} < \frac{4x^{\Theta+1}}{\gamma^2 h},$$

since $(x+h)^{\Theta+1} < (\frac{3}{2}x)^{\Theta+1} \leqslant (\frac{3}{2})^2 x^{\Theta+1} < 3x^{\Theta+1}$; also

$$|w_\rho| = \left|\frac{1}{\rho h}\int_x^{x\pm h} u^{\rho}\,du\right| \leqslant \frac{1}{|\rho|}(x+h)^{\Theta} < \frac{2x^{\Theta}}{|\gamma|}.$$

[1] Holmgren 1; Landau 6, 5–10.

Hence

$$|\sum_\rho w_\rho| < 8x^\Theta \sum_{\gamma>0} \text{Min}\left(\frac{x}{\gamma^2 h}, \frac{1}{\gamma}\right) = 8x^\Theta \sum_{0<\gamma\leqslant x/h} \frac{1}{\gamma} + 8x^\Theta \sum_{\gamma>x/h} \frac{x}{\gamma^2 h}$$

$$= O\left(x^\Theta \log^2 \frac{x}{h}\right) + O\left(x^\Theta \cdot \frac{x}{h} \cdot \frac{h}{x} \log \frac{x}{h}\right) = O(x^\Theta \log^2 x)$$

by Theorem 25 *b* (since $2 < x/h \leqslant x$). Now (cf. p. 64) $\psi(x)$ lies between the two expressions on the left of (32); hence

$$\psi(x) = x + O(h) + O(x^\Theta \log^2 x).$$

Taking, for example, $h = 1$, we deduce (30).

We can obtain the less precise relation $\psi(x) = x + O(x^{\Theta+\epsilon})$ (for any fixed positive ϵ) more directly by a modification of the method of Chapter III, using the line $\sigma = \Theta + \frac{1}{2}\epsilon$ as path of integration. If we work with $\psi_1(x)$, we must use a differencing argument *before* estimating the integral along $\sigma = \Theta + \frac{1}{2}\epsilon$. If we work directly with $\psi(x)$, we take a finite range of integration $(-T, T)$ and make T a function of x, as in the first of the above proofs of (30).

If the Riemann hypothesis is true, Theorem 30 gives

$$\psi(x) = x + O(x^{\frac{1}{2}} \log^2 x), \qquad \pi(x) = \text{li}\, x + O(x^{\frac{1}{2}} \log x).$$

These results, obtained by von Koch[1] in 1901, are the best known of their kind on the Riemann hypothesis.

9. The equation (30) connects the order of magnitude of $\psi(x) - x$ (for large x) with the number Θ, but only in a one-sided way. Now it is easy to establish a relationship in the opposite sense, and we are thus able to conclude that (within certain limits of accuracy) the order of $\psi(x) - x$ is completely determined by Θ. As a measure of the order of $\psi(x) - x$ we introduce a number Θ' defined as follows:

Θ' is the lower bound of the numbers α for which

$$\psi(x) - x = O(x^\alpha). \tag{33}$$

We then have

Theorem 31. $\Theta' = \Theta$.

For in the first place $\Theta' \leqslant \Theta$ by (30). On the other hand, by equation (17), p. 18, we have

$$-\frac{\zeta'}{\zeta}(s) - \frac{s}{s-1} = s\int_1^\infty \frac{\psi(x) - x}{x^{s+1}}\, dx, \tag{34}$$

in the first instance for $\sigma > 1$. Now, if δ is any fixed positive number, we have, by the definition of Θ',

$$\left|\frac{\psi(x) - x}{x^{s+1}}\right| < A(\delta)\frac{x^{\Theta'+\delta}}{x^{\sigma+1}} < A(\delta)x^{-1-\delta} \qquad (\sigma > \Theta' + 2\delta,\ x > 1),$$

[1] von Koch 1.

so that the integral on the right of (34) is uniformly convergent for $\sigma > \Theta' + 2\delta$ and so represents a regular function in the half-plane $\sigma > \Theta'$. It follows from (34) that $\zeta(s)$ cannot have a zero in $\sigma > \Theta'$, or in other words that $\Theta \leq \Theta'$. This completes the proof.

Of the two inequalities $\Theta' \leq \Theta$ and $\Theta \leq \Theta'$ the latter is, of course, the more elementary. The proof is similar in principle to the argument used (in Chapter II, § 9, pp. 36–37) to show that the prime number theorem implies that $\zeta(s)$ has no zeros on the line $\sigma = 1$, but is simpler in detail.

It follows from Theorem 31 and the definition of Θ' that (33) cannot hold for any $\alpha < \Theta$. Since $\Theta \geq \frac{1}{2}$ we conclude that (33) is certainly false if $\alpha < \frac{1}{2}$; and more than this can be asserted if the Riemann hypothesis is false (i.e. if $\Theta > \frac{1}{2}$). Thus we see that the complex zeros of $\zeta(s)$ impose a definite limitation on the degree of accuracy with which $\psi(x)$ can be represented by x, or $\pi(x)$ by $\operatorname{li} x$. The further development of this observation leads to some very remarkable results which will be discussed in detail in the next chapter.

CHAPTER V

IRREGULARITIES OF DISTRIBUTION

1. Up to the present we have been concerned for the most part with the investigation of superior limits to the order of magnitude of the differences $\pi(x) - \operatorname{li} x$, etc., that is to say with theorems which imply a certain degree of regularity in the distribution of primes. Our object in this chapter is to obtain results in the opposite sense. These will enable us to decide the question of the general validity of the inequality $\pi(x) < \operatorname{li} x$ referred to in the Introduction, and it was indeed this problem which gave rise to the deeper theorems of the present chapter.

Our theorems are most conveniently stated in the 'Ω-notation', which forms a natural complement to the classical O- and o-notations. We write

$$f(x) = \Omega(x)$$

(as $x \to \infty$) if there exists a positive number c independent of x such that $|f(x)| > cx$ for arbitrarily large values of x; thus 'Ω' is the negation of 'o'. If $f(x)$ is a real function, we write

$$f(x) = \Omega_{+}(x)$$

if $f(x) > cx$ for arbitrarily large x, and

$$f(x) = \Omega_{-}(x)$$

if $f(x) < -cx$ for arbitrarily large x. Thus 'Ω' is equivalent (for a real f) to 'either Ω_{+} or Ω_{-}'. We shall use the symbol '$\Omega_{\pm}$' to denote 'both Ω_{+} and Ω_{-}'. Thus

$$x \sin x = \Omega_{\pm}(x), \quad x + x \sin x = \Omega(x),$$

but we cannot replace Ω in the second relation by $\Omega_{\pm}$. The special function x on the right is used merely by way of illustration; it may of course be replaced by other positive functions.

The Ω-notation was introduced by Hardy and Littlewood; we have replaced their Ω_R and Ω_L by Ω_{+} and Ω_{-}.[1]

The type of theorem at which we are aiming is illustrated by

[1] Hardy and Littlewood **2**, 138.

the remark at the end of the last chapter (§ 9), which may now be expressed by saying that

$$\psi(x) - x = \Omega(x^{\alpha})$$

for every $\alpha < \Theta$, and so certainly for every $\alpha < \frac{1}{2}$.

It is important, however, to investigate not only the magnitude of the differences $\psi(x) - x$, etc., but also their sign, and our first object is to replace Ω in the relation just written by $\Omega_{\pm}$. For this purpose we require some results in the theory of Dirichlet's series and integrals, of which we give a brief account in the next section.

2. Dirichlet's series and integrals. The theory (in its simplest form) deals with series and integrals of the type

$$(1) \qquad \sum_{1}^{\infty} \frac{c_n}{n^s}, \quad \int_{1}^{\infty} \frac{c(x)}{x^s}\,dx,$$

where $c(x)$ is supposed bounded and integrable (in the sense of Riemann) over any finite interval $1 \leqslant x \leqslant X$, and $s = \sigma + ti$ is a complex variable. The following results are classical.

I. *If a Dirichlet's series (or integral) converges for $s = s_1 = \sigma_1 + t_1 i$, it converges for every $s = \sigma + ti$ with $\sigma > \sigma_1$, and indeed uniformly in any fixed angle of the type*

$$|\arg(s - s_1)| \leqslant \alpha < \tfrac{1}{2}\pi.$$

II. *If a Dirichlet's series (or integral) is convergent for some, but not for all, values of s, there exists a unique real number σ_0 such that the series (or integral) is convergent for all s in the half-plane $\sigma > \sigma_0$ but for no s in the half-plane $\sigma < \sigma_0$.*

DEFINITIONS. *The number σ_0 is called the abscissa of convergence, the line $\sigma = \sigma_0$ the line of convergence, and the half-plane $\sigma > \sigma_0$ the half-plane of convergence, of the Dirichlet's series (or integral).* (In the extreme cases of convergence for all, or for no, values of s we may write $\sigma_0 = -\infty$, or $\sigma_0 = +\infty$.)

III. *A Dirichlet's series (or integral) represents in its half-plane of convergence a regular function of s whose successive derivatives are obtained by differentiation term-by-term (or under the integral sign).*

Of these results I is easily proved by partial summation (or integration), II then follows by a Dedekind section argument, and III is deduced from I on the basis of Weierstrass's theorems (p. 25, footnote).[1]

There is some analogy between Dirichlet's series $\Sigma c_n n^{-s}$ and power series $\Sigma c_n z^n$; and indeed each type of series may be regarded as a special case of the generalised Dirichlet's series

$$(2) \qquad \sum_{1}^{\infty} c_n e^{-\lambda_n s} \qquad (\lambda_1 < \lambda_2 < \ldots, \ \lambda_n \to \infty)$$

if we make the substitution $z = e^{-s}$ in the power series. But the analogy breaks down at a number of important points. For example, the line of convergence of a Dirichlet's series (which corresponds to the circle of convergence of a power series) bears no necessary relation to the singularities of the analytic function $f(s)$ defined (on the basis of III) by the series. This is illustrated by the series $1 - 2^{-s} + 3^{-s} - \ldots$, which has $\sigma_0 = 0$ but represents the function $(1 - 2^{1-s})\,\zeta(s)$ which has no singularities in the finite part of the plane. There is, however, one important special case in which σ_0 is determined by the singularities of $f(s)$.

Theorem H. *If c_n (or $c(x)$), supposed real, is of constant sign for all sufficiently large n (or x), then the real point $s = \sigma_0$ of the line of convergence of the Dirichlet's series (or integral) (1) is a singularity of the function $f(s)$ represented by the series (or integral).*

In case the analytic function $f(s)$ is many-valued the singularity contemplated is one directly associated with the single-valued branch represented by (1).

Suppose the theorem false. Then there exist a domain D containing both the half-plane $\sigma > \sigma_0$ and the point $s = \sigma_0$, and a function $f(s)$ regular in D and identical with (1) in $\sigma > \sigma_0$. Take an $\alpha > \sigma_0$, and choose R so that the domain $|s - \alpha| < R$ contains the point $s = \sigma_0$ but is itself contained in D (so that $R > \alpha - \sigma_0$). Then, since $f(s)$ is regular in D,

$$f(s) = \sum_{0}^{\infty} \frac{(s-\alpha)^n}{n!} f^{(n)}(\alpha) = \sum_{0}^{\infty} \frac{(\alpha - s)^n}{n!} (-1)^n f^{(n)}(\alpha) \quad (|s - \alpha| < R).$$

[1] For details see **HR**, 3–5.

Take for definiteness the integral, and suppose (as we may) that $c(x) \geqslant 0$ for $x > x_1$. Then, since $\alpha > \sigma_0$,

$$(-1)^n f^{(n)}(\alpha) = \int_1^\infty \frac{c(x)(\log x)^n}{x^\alpha}\,dx \geqslant \int_1^X \frac{c(x)(\log x)^n}{x^\alpha}\,dx,$$

if $X > x_1$. Hence, for real s satisfying $\alpha - R < s < \alpha$,

$$f(s) \geqslant \sum_0^\infty \frac{(\alpha - s)^n}{n!}\int_1^X \frac{c(x)(\log x)^n}{x^\alpha}\,dx = \int_1^X \sum_0^\infty \frac{(\alpha-s)^n}{n!}\frac{c(x)(\log x)^n}{x^\alpha}\,dx,$$

the last series being uniformly convergent for $1 < x < X$ since the general term does not exceed $K(R\log X)^n/n!$ in absolute value, where K is the upper bound of $|c(x)x^{-\alpha}|$ in $1 < x < X$. Summing the series under the integral sign, we deduce

$$f(s) \geqslant \int_1^X \frac{c(x)}{x^\alpha} e^{(\alpha - s)\log x}\,dx = \int_1^X \frac{c(x)}{x^s}\,dx.$$

Since the integral on the right is an increasing function of X (for $X > x_1$), it follows that the Dirichlet's integral is convergent at all points s of the segment $\alpha - R < s < \alpha$. But this is impossible, since this segment extends to the left of the line $\sigma = \sigma_0$.

Theorem H is due to Landau. The corresponding theorem for power series had been proved by Pringsheim, but his proof is not applicable to Dirichlet's series.[1]

We may note in passing that this theorem may be used to prove that $\zeta(s)$ has no zeros on the line $\sigma = 1$.[2] We start from the identity

$$\frac{\zeta^2(s)\,\zeta(s+\gamma i)\,\zeta(s-\gamma i)}{\zeta(2s)} = \sum_1^\infty \frac{|C_n|^2}{n^s} \qquad (\sigma > 1), \tag{3}$$

where $\gamma \gtrless 0$, $C_n = \sum_{d\mid n} d^{\gamma i}$, the last sum being over all (positive) divisors d of n; this identity (which is a special case of one due to Ramanujan) may be proved by applying Euler's identity (Theorem 5) to the series on the right. Let σ_0 be the abscissa of convergence of this series. Then $\sigma_0 \leqslant 1$, and (3) is valid (by analytic continuation if $\sigma_0 < 1$) for $\sigma > \sigma_0$, the left-hand side $f(s)$ being of necessity regular in this half-plane. Since $|C_n|^2 \geqslant 0$, it follows from Theorem H that (f being single-valued) σ_0 is determined as the singularity of $f(s)$ on the real axis lying furthest to the right. If $\zeta(s)$ had a zero at $1+\gamma i$ (and therefore another at $1-\gamma i$) this would give $\sigma_0 = -1$, which is easily seen, in various ways, to be impossible; for example, (3) would then give $f(\tfrac12) \geqslant |C_1|^2 = 1$, whereas in fact $f(\tfrac12) = 0$.

[1] Landau 1. See also Landau, *Darstellung und Begründung einiger neuerer Ergebnisse der Funktionentheorie*, 2nd ed. (Berlin, Springer, 1929), 14.

[2] Ingham 1.

3. We now apply the results of § 2 to the proof of

Theorem 32. *If δ is any fixed positive number, then*

$$\psi(x) - x = \Omega_{\pm}(x^{\Theta-\delta}), \tag{4}$$

$$\Pi(x) - \operatorname{li} x = \Omega_{\pm}(x^{\Theta-\delta}). \tag{5}$$

Θ denotes as before the upper bound of the real parts of the zeros of $\zeta(s)$.

We must prove (4) and (5) separately, as we cannot deduce (5) from (4) by partial summation. We give the proof of (5); that of (4) is similar but easier.

Let $0 < \alpha < \Theta$. For real $s > 1$ we have

$$s\int_1^\infty \frac{\Pi(x)}{x^{s+1}}\,dx = \log\zeta(s)$$

by equation (18), p. 18; also

$$s\int_e^\infty \frac{\operatorname{li} x}{x^{s+1}}\,dx = \left[-\frac{\operatorname{li} x}{x^s}\right]_e^\infty + \int_e^\infty \frac{dx}{x^s \log x} = \frac{\operatorname{li} e}{e^s} + \int_{s-1}^\infty \frac{e^{-y}}{y}\,dy \qquad (x^{s-1} = e^y)$$

$$= \frac{\operatorname{li} e}{e^s} + \int_{s-1}^1 \frac{dy}{y} + \int_{s-1}^1 \frac{e^{-y}-1}{y}\,dy + \int_1^\infty \frac{e^{-y}}{y}\,dy$$

$$= -\log(s-1) + g(s),$$

where $g(s)$ is an integral function; and

$$\int_1^\infty \frac{x^\alpha}{x^{s+1}}\,dx = \frac{1}{s-\alpha}.$$

Combining these results, we obtain

$$\int_1^\infty \frac{c(x)}{x^s}\,dx = f(s) \qquad (s > 1), \tag{6}$$

where

$$c(x) = \frac{\Pi(x) - \operatorname{li} x - x^\alpha}{x} \quad (x \geqq e), \qquad \frac{\Pi(x) - x^\alpha}{x} \quad (1 \leqq x < e),$$

$$f(s) = \frac{1}{s}\log\{(s-1)\zeta(s)\} - \frac{1}{s-\alpha} - \frac{g(s)}{s}. \tag{7}$$

Let σ_0 be the abscissa of convergence of the Dirichlet's integral in (6). Then the integral represents a single-valued branch of $f(s)$ regular in $\sigma > \sigma_0$. Now it is clear from (7) that no such branch could exist if the half-plane $\sigma > \sigma_0$ contained any zeros of $\zeta(s)$. Hence we must have $\sigma_0 \geqq \Theta$. On the other hand $f(s)$ has no singularities on the stretch $s > \alpha$ of the real axis, $(s-1)\zeta(s)$ being regular and different from zero along this stretch. In particular, since $\sigma_0 \geqslant \Theta > \alpha$, the point $s = \sigma_0$ is not a singularity of $f(s)$. It follows from Theorem H that we cannot have, for example, $c(x) \leqslant 0$ for all sufficiently large x. Hence $c(x) > 0$, that is $\Pi(x) - \mathrm{li}\, x > x^{\alpha}$, for arbitrarily large x. And we can prove similarly that $\Pi(x) - \mathrm{li}\, x < -x^{\alpha}$ for arbitrarily large x. Since α may be any number satisfying $0 < \alpha < \Theta$, this proves (5).

It will be noted that the argument depends essentially on the fact that $\zeta(s)$ has no zeros on the positive real axis, a property which is not usually important.

Since $\psi(x) - x = \psi([x]) - [x] + O(1)$, (4) is equivalent to

$$\psi(n) - n = \Omega_{\pm}(n^{\Theta-\delta})$$

when $n \to \infty$ by integral values. This implies in particular that, if $W(n)$ is the number of changes of sign in the sequence

$$\psi(1) - 1,\ \psi(2) - 2,\ \ldots,\ \psi(n) - n,$$

then $W(n) \to \infty$ as $n \to \infty$. Pólya has obtained the more precise result

$$\overline{\lim_{n\to\infty}} \frac{W(n)}{\log n} \geqslant \frac{c}{\pi},$$

where c is defined as follows. If $\zeta(s)$ has zeros $\Theta + \gamma i$ on the line $\sigma = \Theta$, then c is the least positive γ corresponding to these zeros; otherwise $c = +\infty$.[1] The proof is based on a refinement of Theorem H.

4. Since $\Theta \geqslant \frac{1}{2}$ it follows from (4) that $\psi(x) - x = \Omega_{\pm}(x^{\frac{1}{2}-\delta})$. By a refinement of the argument we can prove

Theorem 33. $\psi(x) - x = \Omega_{\pm}(x^{\frac{1}{2}})$.

If $\Theta > \frac{1}{2}$ this (and more) follows from (4) since we can choose δ so that $\Theta - \delta > \frac{1}{2}$. We therefore suppose $\Theta = \frac{1}{2}$. We have, for $\sigma > 1$,

$$\int_1^{\infty} \frac{c(x)}{x^s}\, dx = f(s), \tag{8}$$

where $$c(x) = \frac{\psi(x) - x + cx^{\frac{1}{2}}}{x}, \quad f(s) = -\frac{1}{s}\frac{\zeta'}{\zeta}(s) - \frac{1}{s-1} + \frac{c}{s-\frac{1}{2}},$$

c being a positive constant. Suppose, if possible, that $c(x) \geqslant 0$ for all $x \geqslant X (> 1)$. Then, since $f(s)$ has a singularity at $s = \frac{1}{2}$ but at no point on the real axis to the right of this, it follows from Theorem H that $\sigma_0 = \frac{1}{2}$ (σ_0 being the abscissa of convergence of the integral in (8)), so that (8) is valid for $\sigma > \frac{1}{2}$.[2] Hence, for $\sigma > \frac{1}{2}$,

$$|f(\sigma + ti)| \leqslant \int_1^X \frac{|c(x)|}{x^{\sigma}}\, dx + \int_X^{\infty} \frac{c(x)}{x^{\sigma}}\, dx = \int_1^X \frac{|c(x)| - c(x)}{x^{\sigma}}\, dx + f(\sigma)$$

$$\leqslant 2\int_1^X \frac{|c(x)|}{x^{\frac{1}{2}}}\, dx + f(\sigma) = K + f(\sigma),$$

[1] Pólya 1.

[2] Theorem H, which is not used again in the proof, may be dispensed with if desired. For the convergence for $\sigma > \frac{1}{2} (= \Theta)$ of the integral in (8) is secured (independently of any assumption about the sign of $c(x)$) by Theorem 30.

where K is independent of σ and t. Take $t=\gamma_1$, where $\frac{1}{2}+\gamma_1 i$ is the zero with least positive γ, multiply both sides by $\sigma-\frac{1}{2}$, and let $\sigma\to\frac{1}{2}+0$; this gives

$$\frac{m_1}{|\frac{1}{2}+\gamma_1 i|}<c,$$

where m_1 is the order of multiplicity of the zero $\frac{1}{2}+\gamma_1 i$. Now c is at our disposal; choose it so that $0<c<m_1/|\frac{1}{2}+\gamma_1 i|$. Then the supposition that $c(x)\geqslant 0$ for $x\geqslant X$ leads to a contradiction, so that we must have $c(x)<0$ for arbitrarily large x; and a similar result holds for the function $c(x)=\{-\psi(x)+x+cx^{\frac{1}{2}}\}/x$. Since c may be taken as near as we please to $m_1/|\frac{1}{2}+\gamma_1 i|$ we conclude that, when $x\to\infty$,

$$\overline{\lim}\,\frac{\psi(x)-x}{x^{\frac{1}{2}}}\geqslant\delta_1,\quad \underline{\lim}\,\frac{\psi(x)-x}{x^{\frac{1}{2}}}\leqslant-\delta_1,$$

where $\delta_1=m_1/|\frac{1}{2}+\gamma_1 i|$.

Theorems 32 and 33 are due to E. Schmidt. Less precise results pointing in the same direction had been obtained earlier by Phragmén.[1] It is worth while to note that we cannot expect to make further progress by these methods. For the same arguments applied to the formula (16) on p. 32 will show that

$$\psi_1(x)-\tfrac{1}{2}x^2=\Omega_{\pm}(x^{\frac{3}{2}}),$$

and we know on the other hand from Theorem 30 that, if $\Theta=\frac{1}{2}$, $\psi_1(x)-\frac{1}{2}x^2=O(x^{\frac{3}{2}})$.

5. If we write

$$P(x)=\pi(x)-\operatorname{li}x,\quad Q(x)=\Pi(x)-\operatorname{li}x,\quad R(x)=\psi(x)-x,$$

we see from Theorem 32 that $Q(x)$ and $R(x)$ change sign infinitely often as x increases to infinity. But the problem of $P(x)$ is more complicated, since $\pi(x)$ is less directly connected with $\zeta(s)$ than are $\Pi(x)$ and $\psi(x)$. Writing $M=[\log x/\log 2]$, we have

$$Q(x)-P(x)=\sum_{m=2}^{M}\frac{\pi(x^{1/m})}{m}=\tfrac{1}{2}\pi(x^{\frac{1}{2}})+O(Mx^{\frac{1}{3}})\sim\frac{x^{\frac{1}{2}}}{\log x},$$

so that

$$P(x)=\frac{x^{\frac{1}{2}}}{\log x}\left(-1+Q(x)\frac{\log x}{x^{\frac{1}{2}}}+o(1)\right). \tag{9}$$

Now the term -1 on the right indicates that negative values of $P(x)$ may be expected to preponderate, but whether $P(x)$ is *always* negative, as suggested by numerical tables, cannot be decided on the basis of the results so far obtained. If $\Theta>\frac{1}{2}$, then, by Theorem 32,

$$\overline{\underline{\lim}}\,\frac{R(x)}{x^{\frac{1}{2}}}=\pm\infty,\quad \overline{\underline{\lim}}\,Q(x)\frac{\log x}{x^{\frac{1}{2}}}=\pm\infty, \tag{10}$$

[1] Phragmén 1, 2; Schmidt 1. For another proof of Theorem 33 see Littlewood 4.

when $x \to \infty$, and the second result combined with (9) shows that $P(x)$ changes sign infinitely often. But the argument fails if $\Theta = \frac{1}{2}$.[1]

In order to decide the question of the variations of sign of $P(x)$ we shall prove a further set of theorems which will show that the relations (10) are true whether the Riemann hypothesis is true or false. These theorems are much more difficult, and it will be convenient to indicate in outline the general idea of the argument before going into detail. Starting from the explicit formula for $\psi_0(x)$ (Theorem 29) and assuming the Riemann hypothesis (as we may), we replace the denominators $\rho = \frac{1}{2} + \gamma i$ in $\Sigma x^\rho/\rho$ by γi (as we may with a sufficient approximation), and the series itself therefore by

$$2x^{\frac{1}{2}} \sum_{\gamma > 0} \frac{\sin(\gamma \log x)}{\gamma} = 2x^{\frac{1}{2}} \sum_{n=1}^{\infty} \frac{\sin(\gamma_n \log x)}{\gamma_n}.$$

Now the last series (without the factor $2x^{\frac{1}{2}}$) is, formally, $-\mathfrak{I} G(i \log x)$, where

$$G(s) = G(\sigma + ti) = \sum_{1}^{\infty} \frac{e^{-\gamma_n s}}{\gamma_n}, \tag{11}$$

and we wish to show that, when $t \to \infty$,

$$\overline{\lim}\, \mathfrak{I} G(ti) = +\infty, \quad \underline{\lim}\, \mathfrak{I} G(ti) = -\infty. \tag{12}$$

We begin by investigating $G(s)$ for $\sigma > 0$. The series defining $G(s)$ is absolutely and uniformly convergent in any fixed half-plane $\sigma > \delta > 0$ since

$$\left| \frac{e^{-\gamma_n s}}{\gamma_n} \right| = \frac{1}{\gamma_n e^{\gamma_n \sigma}} < \frac{1}{\gamma_n \cdot \gamma_n \sigma} < \frac{1}{\gamma_n^2 \delta},$$

and $\Sigma 1/(\gamma_n^2 \delta)$ is convergent; thus $G(s)$ is regular in $\sigma > 0$. We first show, by means of a theorem of Dirichlet on Diophantine approximation, that there exist points $s = \sigma + ti$ with small positive σ and large positive t at which $\mathfrak{I} G(s)$ is large and positive, and other points at which $\mathfrak{I} G(s)$ is large and negative.

[1] By the method of § 4 we could show that

$$\overline{\lim}\, Q(x) \frac{\log x}{x^{\frac{1}{2}}} \geqslant \delta_1 = \frac{m_1}{|\frac{1}{2} + \gamma_1 i|},$$

and this would suffice to show that $P(x)$ changes sign infinitely often if δ_1 were greater than 1. But actually δ_1 is about 0·07. (See T, 45–47.)

But these points, though 'near' the line $\sigma = 0$, are not sufficiently near to enable us to make trivial deductions about the behaviour of $\mathfrak{J}G(s)$ *on* this line, and to make the transition we use a theorem of general function-theory due to Phragmén and Lindelöf. When we come to details, we find that we have to use one of the inequalities for the 'remainder' $R(x, T)$ in Theorem 29; also we do not actually define $G(s)$ on the line $\sigma = 0$.

6. We begin by proving the general theorems of Dirichlet and Phragmén-Lindelöf referred to in the preceding section.

Theorem J. *Let $\theta_1, \theta_2, \ldots, \theta_N$ be N real numbers, and q a positive integer. Then in every interval of the form*

$$\tau \leqslant t \leqslant \tau q^N \qquad (\tau > 0)$$

there exists a number t such that each of the products

$$t\theta_1, t\theta_2, \ldots, t\theta_N$$

differs from an integer by less than $1/q$.

We use (merely for convenience) the language of N-dimensional geometry. Consider the 'unit cube' in N dimensions, i.e. the set of 'points' $\{x_1, x_2, \ldots, x_N\}$ defined by

$$0 \leqslant x_1 < 1, \ldots, 0 \leqslant x_N < 1,$$

and subdivide it into q^N small cubes by 'planes' parallel to the coordinate planes, a typical small cube being defined by

$$\frac{n_1 - 1}{q} \leqslant x_1 < \frac{n_1}{q}, \ldots, \frac{n_N - 1}{q} \leqslant x_N < \frac{n_N}{q},$$

where $n_1, n_2, \ldots, n_N$ are positive integers not greater than q.

Let τ be a positive number, and consider the points

$$P_r = \{(r\tau\theta_1), (r\tau\theta_2), \ldots, (r\tau\theta_N)\} \qquad (r = 0, 1, 2, \ldots),$$

where $(u) = u - [u]$. Each P_r lies in the unit cube and therefore in some small cube. Since there are only q^N small cubes, one at least must contain more than one of the $q^N + 1$ points $P_0, P_1, \ldots, P_{q^N}$. Suppose P_r and P_s $(0 \leqslant s < r \leqslant q^N)$ lie in the same small cube. Then

$$|(r\tau\theta_n) - (s\tau\theta_n)| < 1/q \qquad (n = 1, 2, \ldots, N),$$

i.e.

$$|h\tau\theta_n - k_n| < 1/q \qquad (n = 1, 2, \ldots, N),$$

where $h = r - s$ and $k_n = [r\tau\theta_n] - [s\tau\theta_n]$. Since $1 < h < q^N$ and k_n is an integer, the number $t = h\tau$ fulfils our requirements.

We observe that t may be taken to be an integral multiple of τ, but this is irrelevant to the application.

Theorem K. *Let D be the domain of the s-plane defined by*

$$t > t_0, \quad g_1(t) < \sigma < g_2(t),$$

where $g_1(t)$ and $g_2(t)$ are continuous functions satisfying

$$\alpha_1 \leqslant g_1(t) < g_2(t) \leqslant \alpha_2 \qquad (t \geqslant t_0),$$

α_1 and α_2 being constants. Suppose that $f(s)$ is regular in D and continuous in D' (the region consisting of D and its boundary), and that it satisfies throughout D' an inequality of the type

$$(13) \qquad |f(s)| < Ke^{e^{ct}}, \text{ where } 0 < c < \frac{\pi}{\alpha_2 - \alpha_1}$$

(K and c being constants). Then, if the inequality

$$(14) \qquad |f(s)| \leqslant C$$

(C a constant) holds at all points on the boundary of D, it must hold at all points inside D.

We may suppose $\alpha_1 = -\frac{1}{2}\pi$, $\alpha_2 = \frac{1}{2}\pi$, since this can be secured by the linear transformation $s' = \pi\{s - \frac{1}{2}(\alpha_1 + \alpha_2)\}/(\alpha_2 - \alpha_1)$; in this case $0 < c < 1$.

Suppose, if possible, that (14) is true for all boundary points but false for at least one interior point $s^* = \sigma^* + t^*i$; so that $|f(s^*)| = C + 2\delta$, where $\delta > 0$. Consider the function

$$\phi(s) = f(s)\, e^{-\epsilon \cos bs},$$

where $c < b < 1$ and ϵ is a positive number chosen so small that

$$(15) \qquad |\phi(s^*)| > C + \delta.$$

This is clearly possible (ϵ depending, of course, on s^*). In D' we have, since $0 < b < 1$ and $-\frac{1}{2}\pi \leqslant \sigma \leqslant \frac{1}{2}\pi$,

$$|\phi(s)| = |f(s)|\, e^{-\epsilon \cos b\sigma \cosh bt} \leqslant |f(s)|\, e^{-\frac{1}{2}\epsilon \cos \frac{1}{2}b\pi . e^{bt}} \leqslant |f(s)|.$$

Hence $|\phi(s)| \leqslant C$ on the boundary of D, and

$$|\phi(s)| < K \exp(e^{ct} - \tfrac{1}{2}\epsilon \cos \tfrac{1}{2}b\pi . e^{bt})$$

inside. Since $\frac{1}{2}\epsilon \cos \frac{1}{2}b\pi > 0$ and $b > c$, the expression on the right tends to 0 when $t \to \infty$. Hence we can find a $T > t^*$ so that this expression is less than $C + \delta$ when $t = T$. It follows that, if D_T is the (bounded) domain defined by $t_0 < t < T$, $g_1(t) < \sigma < g_2(t)$, then $|\phi(s)| < C + \delta$ on the whole of the boundary of D_T. Hence, by the maximum modulus principle (ϕ being regular in D_T and continuous in D_T'), $|\phi(s)| < C + \delta$ at all points inside D_T. But this contradicts (15) since, by our choice of T, s^* lies in D_T.

Theorem K belongs to a fairly wide class of theorems with conclusions of the form 'f is either bounded or of very high order'. It asserts that, if $|f(s)|$ satisfies the inequality (14) on the boundary of D, then it either satisfies the same inequality inside or is of sufficiently high order to violate any inequality of the type (13). That the latter alternative is in fact possible may be seen from the example

$$f(s) = e^{\cos s}, \quad t_0 = 0, \quad g_1(t) = \alpha_1 = -\tfrac{1}{2}\pi, \quad g_2(t) = \alpha_2 = \tfrac{1}{2}\pi,$$

for in this case $|f(s)| = e^{\cos \sigma \cosh t}$, so that $|f(s)| \leqslant e$ on the boundary of D but satisfies no inequality of the type (13) throughout the interior. As, however, a similar inequality with $c = \pi/(\alpha_2 - \alpha_1) = 1$ does hold, the example shows also that the theorem cannot be substantially improved by a relaxation of the condition (13).

Subject to the necessity for some general restriction, such as (13), on the order of magnitude of the function, Theorem K may be regarded as an extension of the maximum modulus principle to a certain class of unbounded domains.

7. We shall now set out in a series of lemmas the relevant properties of the function

$$G(s) = \sum_{1}^{\infty} \frac{e^{-\gamma_n s}}{\gamma_n} = \sum_{\gamma > 0} \frac{e^{-\gamma s}}{\gamma} \qquad (\sigma > 0)$$

introduced in § 5.

Lemma 1. *To every positive ϵ corresponds a positive $\sigma_1 = \sigma_1(\epsilon)$ such that, if $0 < \sigma < \sigma_1$, $T_0 > 0$, the interval*

$$T_0 < T < T_0 e^{(1/\sigma)^{1+\epsilon}} \tag{16}$$

contains a number $T = T(\epsilon, \sigma, T_0)$ with the property that

$$|G(\sigma + ti + Ti) - G(\sigma + ti)| < \epsilon$$

for all real values of t.

We have, for $\sigma > 0$, t, T real,

$$|G(\sigma+ti+Ti)-G(\sigma+ti)|$$

$$=\left|\sum_{\gamma>0}\frac{e^{-\gamma\sigma-\gamma ti-\frac{1}{2}\gamma Ti}(e^{-\frac{1}{2}\gamma Ti}-e^{\frac{1}{2}\gamma Ti})}{\gamma}\right|<2\sum_{\gamma>0}\frac{e^{-\gamma\sigma}|\sin\frac{1}{2}\gamma T|}{\gamma}$$

$$<2\sum_{\gamma>U}\frac{e^{-\gamma\sigma}}{\gamma}+2\sum_{n=1}^{N(U)}\frac{|\sin\frac{1}{2}\gamma_n T|}{\gamma_n}=2\Sigma_1+2\Sigma_2,$$

say, where $U>3$ and $N(U)$ denotes as usual the number of zeros with $0<\gamma<U$. Since $e^x>x$ for $x>0$, we have

$$\Sigma_1<\sum_{\gamma>U}\frac{1}{\sigma\gamma^2}<\frac{A_1}{\sigma}\frac{\log U}{U}$$

by Theorem 25 *b*. Next, taking a positive integer q and applying Theorem J to the numbers $\theta_n=\gamma_n/2\pi$ $[n=1, 2, \ldots, N(U)]$, we see that there exists a T in the interval

$$T_0<T<T_0 q^{N(U)} \tag{17}$$

such that

$$\frac{T\gamma_n}{2\pi}=k_n+\phi_n,\quad |\phi_n|<\frac{1}{q},\qquad [n=1, 2, \ldots, N(U)],$$

the k_n being integers. If T is so chosen, we have

$$\Sigma_2=\sum_{n=1}^{N(U)}\frac{|\sin\pi\phi_n|}{\gamma_n}<\sum_{1}^{N(U)}\frac{|\pi\phi_n|}{\gamma_n}<\frac{\pi}{q}\sum_{0<\gamma\leqslant U}\frac{1}{\gamma}<\frac{A_2\log^2 U}{q}$$

by Theorem 25 *b*, and therefore

$$|G(\sigma+ti+Ti)-G(\sigma+ti)|<A_3\left(\frac{\log U}{\sigma U}+\frac{\log^2 U}{q}\right)$$

for $\sigma>0$ and all real t. In this take

$$U=\frac{B}{\sigma}\log\frac{1}{\sigma},\quad q=\left[B\log^2\frac{1}{\sigma}\right]+1,\quad B=\frac{6A_3}{\epsilon},$$

assuming that $0<\sigma<\sigma_0$, where $\sigma_0=\sigma_0(\epsilon)$ is chosen so small that $3<U<(1/\sigma)^2$ for $0<\sigma<\sigma_0$. Then

$$|G(\sigma+ti+Ti)-G(\sigma+ti)|<A_3\left(\frac{2\log(1/\sigma)}{B\log(1/\sigma)}+\frac{4\log^2(1/\sigma)}{B\log^2(1/\sigma)}\right)=\epsilon.$$

Also, since $N(U) = O(U \log U)$ by Theorem 25, we have

$$N(U)\log q < N\left(\frac{B}{\sigma}\log\frac{1}{\sigma}\right).\log\left(B\log^2\frac{1}{\sigma}+1\right) < \left(\frac{1}{\sigma}\right)^{1+\epsilon}$$

if $0 < \sigma < \sigma_1 = \sigma_1(B, \epsilon) = \sigma_1(\epsilon) < \sigma_0$. Thus (17) implies (16) if $0 < \sigma < \sigma_1$, and the lemma is proved.

Lemma 2. *If* $s = \sigma + ti = re^{\theta i}$ $(r > 0, -\frac{1}{2}\pi < \theta < \frac{1}{2}\pi)$, *then*

$$\mathfrak{I}G(s) = -\frac{\theta}{2\pi}\log\frac{1}{\sigma} + O(1),$$

when $r \to 0$ *(θ fixed).*

We have

$$G(s) = s\int_{\gamma_1}^{\infty} C(x)e^{-sx}dx \qquad (\sigma > 0), \tag{18}$$

where
$$C(x) = \sum_{0<\gamma\leqslant x}\frac{1}{\gamma}.$$

This follows from Theorem A (p. 18), with $\lambda_n = \gamma_n$, $c_n = 1/\gamma_n$, $\phi(x) = e^{-sx}$, since $C(X)e^{-sX} = O\{N(X)e^{-\sigma X}\} = o(1)$ as $X \to \infty$. Now, for $x > \gamma_1$,

$$C(x) = \int_{\gamma_1}^{x}\frac{N(u)}{u^2}du + \frac{N(x)}{x}$$

(again by Theorem A); whence by Theorem 25

$$2\pi C(x) = \tfrac{1}{2}(\log x)^2 + B\log x + O(1)$$

as $x \to \infty$, where B is a real constant. Substituting into (18) we deduce that

$$2\pi G(s) = \tfrac{1}{2}I_2 + BI_1 + R, \tag{19}$$

where
$$I_k = s\int_{\gamma_1}^{\infty}(\log x)^k e^{-sx}dx \qquad (k = 1, 2),$$

$$|R| < |s|\int_{\gamma_1}^{\infty}Ae^{-\sigma x}dx < A\frac{|s|}{\sigma} = \frac{A}{\cos\theta}.$$

Let I_k' be I_k with the lower limit γ_1 replaced by 0. Then

$$|I_k - I_k'| = \left|s\int_0^{\gamma_1}(\log x)^k e^{-sx}dx\right| < |s|\int_0^{\gamma_1}|\log x|^k dx = A_k r.$$

Now, by the substitution $sx = y$ and an application of Cauchy's theorem,

$$I_k' = \int_0^\infty \left(\log \frac{y}{s}\right)^k e^{-y} dy = \int_0^\infty \left(\log \frac{1}{s} + \log y\right)^k e^{-y} dy.$$

Expanding $(\ldots)^k$ and integrating term-by-term we see that I_k' is a polynomial of degree k in $\log 1/s$ whose leading term is $(\log 1/s)^k$. Substituting these results into (19) we obtain

$$2\pi G(s) = \tfrac{1}{2}\left(\log \frac{1}{s}\right)^2 + B' \log \frac{1}{s} + O(1)$$

as $r \to 0$, B' being a real constant. Since

$$\log \frac{1}{s} = \log \frac{1}{r} - \theta i = \log \frac{1}{\sigma} + \log \cos \theta - \theta i,$$

it follows that, when $r \to 0$,

$$2\pi \Im G(s) = -\theta \log \frac{1}{r} - B'\theta + O(1) = -\theta \log \frac{1}{\sigma} + O(1).$$

Lemma 3. *If* $0 < c < \frac{1}{4}$, *then for every sufficiently small positive* σ, *i.e. for* $0 < \sigma < \sigma_0 = \sigma_0(c)$, *there exist a* $t' = t'(c, \sigma)$ *and a* $t'' = t''(c, \sigma)$ *such that*

$$t' > e^{1/\sigma}, \quad \Im G(\sigma + t'i) > + c \log \log t',$$
$$t'' > e^{1/\sigma}, \quad \Im G(\sigma + t''i) < - c \log \log t''.$$

Let $0 < \epsilon < \frac{1}{3}$, and choose $\sigma_1 = \sigma_1(\epsilon)$ in accordance with Lemma 1. Taking $T_0 = e^{(1/\sigma)^{1+\epsilon}}$, we infer from the lemma that for each σ in $0 < \sigma < \sigma_1$ there exists a $T_\sigma = T_\sigma(\epsilon)$ such that

$$e^{(1/\sigma)^{1+\epsilon}} < T_\sigma < e^{2(1/\sigma)^{1+\epsilon}}, \tag{20}$$

$$|G(\sigma + ti + T_\sigma i) - G(\sigma + ti)| < \epsilon \qquad \text{(all real } t\text{)}. \tag{21}$$

Now by Lemma 2, with $\theta = \pm \frac{1}{2}\pi(1 - \epsilon)$, we have [since $s = \sigma(1 + i \tan \theta)$]

$$\mp \Im G(\sigma \pm \lambda \sigma i) \sim \tfrac{1}{4}(1 - \epsilon) \log \frac{1}{\sigma} \qquad (\sigma \to +0),$$

where $\lambda = \tan\{\frac{1}{2}\pi(1 - \epsilon)\}$. Hence, for $0 < \sigma < \sigma_2 = \sigma_2(\epsilon)$,

$$\mp \Im G(\sigma \pm \lambda \sigma i) > \tfrac{1}{4}(1 - 2\epsilon) \log \frac{1}{\sigma}. \tag{22}$$

Taking $t = \pm\lambda\sigma$ in (21) and combining with (22) we obtain

$$\mp\Im G(\sigma \pm \lambda\sigma i + T_\sigma i) > \tfrac{1}{4}(1-2\epsilon)\log\frac{1}{\sigma} - \epsilon > \tfrac{1}{4}(1-3\epsilon)\log\frac{1}{\sigma}$$

for $0 < \sigma < \sigma_3 = \sigma_3(\epsilon)$.

Now it is clear from (20) that, if $0 < \sigma < \sigma_4 = \sigma_4(\epsilon)$,

$$e^{1/\sigma} < \pm\lambda\sigma + T_\sigma < e^{(1/\sigma)^{1+2\epsilon}}.$$

Writing $t' = T_\sigma - \lambda\sigma$ and $t'' = T_\sigma + \lambda\sigma$, we deduce that, if $0 < \sigma < \sigma_5 = \mathrm{Min}(\sigma_3, \sigma_4, \tfrac{1}{2})$, then $t' > e^{1/\sigma}$, $t'' > e^{1/\sigma}$, and

$$\Im G(\sigma + t'i) > \tfrac{1}{4}(1-3\epsilon)\log\frac{1}{\sigma} > \frac{1-3\epsilon}{4(1+2\epsilon)}\log\log\log t' > 0,$$

with similar inequalities for $-\Im G(\sigma + t''i)$. This proves the lemma; for, since $c < \tfrac{1}{4}$, we can choose $\epsilon = \epsilon(c)$ so that the coefficient of $\log\log t'$ is greater than c.

Lemma 4. *If* $0 < \sigma < 1$, *then*

$$|G(s)| < A\left(\log\frac{1}{\sigma} + 1\right)^2.$$

Let $U = e/\sigma \geqslant e$. Then we have, using Theorem 25 *b*,

$$|G(s)| < \sum_{\gamma>0}\frac{e^{-\gamma\sigma}}{\gamma} < \sum_{0<\gamma\leqslant U}\frac{1}{\gamma} + \sum_{\gamma>U}\frac{1}{\sigma\gamma^2} < A_1\log^2 U + A_2\frac{\log U}{\sigma U},$$

whence the result, since $\sigma U = e$, $\log U = \log(1/\sigma) + 1$.

8. We can now state and prove our main theorems.

Theorem 34. *We have*

$$\psi(x) - x = \Omega_\pm(x^{\frac{1}{2}}\log\log\log x)$$

when $x \to \infty$.

We shall prove in fact that when $x \to \infty$

$$(23)\quad \overline{\lim}\,\frac{\psi(x) - x}{x^{\frac{1}{2}}\log\log\log x} > \tfrac{1}{2}, \quad \underline{\lim}\,\frac{\psi(x) - x}{x^{\frac{1}{2}}\log\log\log x} < -\tfrac{1}{2}.$$

If the Riemann hypothesis is false ($\Theta > \tfrac{1}{2}$) these results (and more) follow from Theorem 32. We therefore suppose the Riemann hypothesis true ($\Theta = \tfrac{1}{2}$) in what follows.

By Theorem 29 (16), p. 77, we have

$$\psi_0(x) = x - \sum_{|\gamma| \leqslant T} \frac{x^{\frac{1}{2}+\gamma i}}{\frac{1}{2}+\gamma i} - \frac{\zeta'}{\zeta}(0) - \tfrac{1}{2}\log\left(1 - \frac{1}{x^2}\right) - R(x, T),$$

where

$$|R(x, T)| < A_1\left(x^2 \frac{\log^2 T}{T} + \log x\right) \qquad (x > 2,\ T > 3).$$

In particular (since $(\log^2 u)/u$ decreases for $u > e^2$)

$$|R(x, T)| < A_1\left(x^2 \frac{\log^2 x^2}{x^2} + \log x\right) < A_2 \log^2 x \qquad (x > e,\ T > x^2).$$

Now

$$\left|\sum_{|\gamma| \leqslant T} \frac{x^{\frac{1}{2}+\gamma i}}{\frac{1}{2}+\gamma i} - \sum_{|\gamma| \leqslant T} \frac{x^{\frac{1}{2}+\gamma i}}{\gamma i}\right| = \left|\sum_{|\gamma| \leqslant T} \frac{\frac{1}{2}x^{\frac{1}{2}+\gamma i}}{(\frac{1}{2}+\gamma i)\gamma i}\right| < \sum_{|\gamma| \leqslant T} \frac{x^{\frac{1}{2}}}{2\gamma^2} < A_3 x^{\frac{1}{2}},$$

since $\Sigma \gamma^{-2}$ is convergent. Hence

$$\begin{aligned}\psi(x) - x &= -\sum_{|\gamma| \leqslant T} \frac{x^{\frac{1}{2}+\gamma i}}{\gamma i} + R_1(x, T) \\ &= -2x^{\frac{1}{2}} \sum_{0 < \gamma \leqslant T} \frac{\sin(\gamma \log x)}{\gamma} + R_1(x, T),\end{aligned}$$

where $\qquad |R_1(x, T)| < A_4 x^{\frac{1}{2}} \qquad (x > e,\ T > x^2).$

Putting $x = e^t$, and writing

$$H(t) = \frac{\psi(x) - x}{2x^{\frac{1}{2}}} = \frac{\psi(e^t) - e^t}{2e^{\frac{1}{2}t}}, \quad S_T(t) = -\sum_{0 < \gamma \leqslant T} \frac{\sin \gamma t}{\gamma},$$

we deduce that

$$|H(t) - S_T(t)| < \tfrac{1}{2}A_4 \qquad (t > 1,\ T > e^{2t}). \tag{24}$$

This implies in particular that

$$|S_{T_1}(t) - S_{T_2}(t)| < A_4 \qquad (t > 1,\ T_1 > T_2 > e^{2t}). \tag{25}$$

Now let $G(s) = G(\sigma + ti)$ be the function discussed in § 7. Then, for $\sigma > 0$,

$$\begin{aligned}\Im G(\sigma + ti) - S_T(t) &= \sum_{0 < \gamma \leqslant T} \frac{1 - e^{-\gamma\sigma}}{\gamma} \sin \gamma t - \sum_{\gamma > T} e^{-\gamma\sigma} \frac{\sin \gamma t}{\gamma} \\ &= \Sigma_1 - \Sigma_2,\end{aligned}$$

say. Since $0 < 1 - e^{-u} < u$ for $u > 0$, we have

$$|\Sigma_1| < \sum_{0 < \gamma \leqslant T} \frac{\gamma\sigma}{\gamma} = N(T)\sigma < A_5 T^{\frac{3}{2}}\sigma.$$

Also by Theorem A (p. 18), with

$\lambda_n = \gamma_n$, $c_n = (\sin \gamma_n t)/\gamma_n$ $(\gamma_n > T)$, $c_n = 0$ $(\gamma_n < T)$, $\phi(u) = e^{-u\sigma}$,

we have

$$|\Sigma_2| = \left| \int_T^\infty \sigma e^{-u\sigma} \{S_u(t) - S_T(t)\} du \right| < \int_T^\infty \sigma e^{-u\sigma} A_4 du < A_4$$

by (25), if $t > 1$, $T \geqslant e^{2t}$. Thus

$$|\Im G(\sigma + ti) - S_T(t)| < A_5 T^{\frac{3}{2}} \sigma + A_4 \qquad (t > 1,\ T \geqslant e^{2t}).$$

Combining with (24), taking $T = e^{2t}$, and restricting the range of σ, we obtain

$$(26) \quad |H(t) - \Im G(\sigma + ti)| < A_6 \qquad (t > 1,\ 0 < \sigma < e^{-3t}).$$

Now suppose the first of the inequalities (23) false. Then

$$\overline{\lim_{t\to\infty}} \frac{H(t)}{\log \log t} < \tfrac{1}{4};$$

hence we can find a and t_1, such that $0 < a < \frac{1}{4}$, $t_1 > 3$, and

$$(27) \qquad H(t) < a \log \log t \qquad (t > t_1).$$

Choose b and c so that $a < b < c < \frac{1}{4}$; in what follows the numbers $t_2, t_3, \ldots$ depend on a, b, c, and are supposed chosen so that $t_1 < t_2 < t_3 < \ldots$. By (26) and (27)

$$(28) \qquad \Im G(\sigma + ti) < a \log \log t + A_6 < b \log \log t \qquad (t > t_2,\ 0 < \sigma < e^{-3t}).$$

We now apply Theorem K to the function

$$f(s) = \frac{e^{-iG(s)/b}}{\log s}$$

(where $\log s$ has its principal value) and the region D' defined by $t > t_2$, $e^{-3t} < \sigma < 1$. On the curve $\sigma = e^{-3t}$ $(t > t_2)$, which bounds the region on the left, we have by (28)

$$|f(s)| = \frac{e^{\Im G(s)/b}}{|\log s|} < \frac{e^{\log \log t}}{\log t} = 1.$$

On the remaining part of the boundary $f(s)$ is bounded since $|G(s)| < G(\sigma) < G(e^{-3t_2})$. Hence

$$(29) \qquad |f(s)| < C$$

on the whole boundary, where $C = C(t_2, b) = C(a, b)$. Further

$$|f(s)| < C_1 e^{|G(s)|/b} < C_1 e^{C_2 t^2}$$

throughout D', by Lemma 4, since $\log 1/\sigma < 3t$. Thus the conditions of Theorem K are satisfied, and (29) must therefore hold throughout D'. It follows that

$$e^{\Im G(s)/b} = |f(s) \log s| < C|\log s| \qquad (t > t_2,\ e^{-3t} < \sigma < 1)$$
$$< 2C \log t \qquad (t > t_3,\ e^{-3t} < \sigma < 1).$$

Hence

$$\Im G(s) < b \log (2C \log t) < c \log\log t \qquad (t > t_4,\ e^{-3t} < \sigma < 1).$$

This and (28) together give

$$\text{(30)} \qquad \Im G(\sigma + ti) < c \log\log t \qquad (t > t_4,\ 0 < \sigma < 1).$$

But, since $0 < c < \frac{1}{4}$, this is impossible; for when σ is small enough the point $\sigma + t'i$ of Lemma 3 lies in the region $t > t_4$, $0 < \sigma < 1$, and violates (30). This contradiction establishes the first of the relations (23), and the second is deduced in a similar way from the other half of Lemma 3.

9. We have now to deduce from Theorem 34 the corresponding results for $\Pi(x)$ and $\pi(x)$. The deduction is not entirely trivial, as we are dealing with one-sided inequalities.

Theorem 35. *When* $x \to \infty$,

$$\text{(31)} \qquad \Pi(x) - \operatorname{li} x = \Omega_{\pm}\left(\frac{x^{\frac{1}{2}}}{\log x} \log\log\log x\right),$$

$$\text{(32)} \qquad \pi(x) - \operatorname{li} x = \Omega_{\pm}\left(\frac{x^{\frac{1}{2}}}{\log x} \log\log\log x\right).$$

Suppose first $\Theta = \frac{1}{2}$. Writing (as in § 5)

$$\pi(x) - \operatorname{li} x = P(x), \quad \Pi(x) - \operatorname{li} x = Q(x), \quad \psi(x) - x = R(x),$$

we have, by equation (38), p. 64,

$$Q(x) = \frac{R(x)}{\log x} + \int_2^x \frac{R(u)}{u \log^2 u}\, du + O(1).$$

Hence by partial integration

$$(33)\quad Q(x)-\frac{R(x)}{\log x}=\frac{R_1(x)}{x\log^2 x}-\int_2^x R_1(u)\frac{d}{du}\left(\frac{1}{u\log^2 u}\right)du+O(1),$$

where $$R_1(x)=\int_0^x R(u)\,du=\psi_1(x)-\tfrac{1}{2}x^2=O(x^{\frac{3}{2}})$$

by Theorem 30 (with $\Theta=\frac{1}{2}$). It follows that, if $x>2$,

$$\left|Q(x)-\frac{R(x)}{\log x}\right|<\frac{A_1x^{\frac{3}{2}}}{\log^2 x}+\int_2^x A_1u^{\frac{3}{2}}\left|\frac{d}{du}\left(\frac{1}{u\log^2 u}\right)\right|du+A_2$$

$$<\frac{A_3x^{\frac{1}{2}}}{\log^2 x}+A_4\int_2^x\frac{du}{u^{\frac{1}{2}}\log^2 u}=\frac{A_3x^{\frac{1}{2}}}{\log^2 x}+A_4\int_2^x\frac{u^{\frac{1}{4}}}{\log^2 u}\frac{du}{u^{\frac{3}{4}}}$$

$$<\frac{A_3x^{\frac{1}{2}}}{\log^2 x}+A_4\frac{x^{\frac{1}{4}}}{\log^2 x}\int_2^x\frac{du}{u^{\frac{3}{4}}}<A_5\frac{x^{\frac{1}{2}}}{\log^2 x}$$

if $x>x_0$, since $u^{\frac{1}{4}}/\log^2 u$ ultimately increases and tends to infinity. We infer that

$$(34)\quad \frac{Q(x)}{x^{\frac{1}{2}}(\log x)^{-1}\log\log\log x}-\frac{R(x)}{x^{\frac{1}{2}}\log\log\log x}$$
$$=O\left(\frac{1}{\log x\log\log\log x}\right)=o(1),$$

so that the two quotients on the left have the same limits of indetermination when $x\to\infty$. Thus (31) follows from Theorem 34 if $\Theta=\frac{1}{2}$. And if $\Theta>\frac{1}{2}$ it follows from Theorem 32 (5). In either case (32) follows from (31), since

$$\Pi(x)-\pi(x)=\tfrac{1}{2}\pi(x^{\frac{1}{2}})+O(x^{\frac{1}{3}}\log x)=O\left(\frac{x^{\frac{1}{2}}}{\log x}\right).$$

Theorems 34 and 35 are due to Littlewood.[1] The idea of applying the theory of Diophantine approximation to questions of this kind originates with H. Bohr, who has made numerous applications of the first importance to the general theory of Dirichlet's series.[2] The property of $G(\sigma+ti)$ embodied in Lemma 1 is a kind of 'approximate periodicity' (with respect to t), and it is a somewhat analogous property which forms the basis of the theory of 'almost periodic functions' initiated by Bohr.

10. Theorems 34 and 35 evidently solve the problem which gave rise to them—the problem of the sign of $P(x)=\pi(x)-\operatorname{li}x$. The relation (32) shows that $P(x)$, like $Q(x)$ and $R(x)$, changes

[1] Littlewood 1; Hardy and Littlewood 2.

[2] See T, Chapters I and IV.

sign infinitely often as x increases to infinity. But it is interesting to examine the matter a little more closely from the numerical point of view. By (23) and (34)

$$\overline{\lim} \frac{Q(x) \log x}{x^{\frac{1}{2}} \log\log\log x} > \tfrac{1}{2}.$$

Hence, if $0 < \epsilon < 1$, we have by (9)

$$P(x) > \frac{x^{\frac{1}{2}}}{\log x} \{-1 + \tfrac{1}{2}(1-\epsilon) \log\log\log x - \epsilon\}$$

for arbitrarily large x. Now, although the right-hand side is certainly positive when x is large enough, it remains negative at any rate over the range[1] $10 \leqslant x \leqslant 10^{700}$. Thus, if the above results represent anything approaching the ultimate truth, it is not surprising that no value of x for which $P(x) > 0$ should present itself within the limits of existing tables. How far we must in fact go before reaching such a value is not known; the theoretical reasoning provides no numerical solution of the inequality $P(x) > 0$.

Considerable importance was attached formerly to a function suggested by Riemann as an approximation to $\pi(x)$. If we 'invert' the formula $\Pi(x) = \pi(x) + \frac{1}{2}\pi(x^{\frac{1}{2}}) + \dots$ by a theorem of Möbius[2], we obtain

$$\pi(x) = \Pi(x) - \tfrac{1}{2}\Pi(x^{\frac{1}{2}}) - \tfrac{1}{3}\Pi(x^{\frac{1}{3}}) - \tfrac{1}{5}\Pi(x^{\frac{1}{5}}) + \tfrac{1}{6}\Pi(x^{\frac{1}{6}}) - \dots,$$

the coefficients being the terms of the series (25) on p. 39; and Riemann's function is obtained by replacing each Π by li. This function represents $\pi(x)$ with astonishing accuracy for all values of x for which $\pi(x)$ has been calculated[3], but we now see that its superiority over the function li x is illusory. For the terms $-\frac{1}{2}\operatorname{li} x^{\frac{1}{2}} - \frac{1}{3}\operatorname{li} x^{\frac{1}{3}} - \dots$, which contribute only $O(x^{\frac{1}{2}}/\log x)$, have no influence on formulae of the type of (32); and for special values of x (as large as we please) the one approximation will deviate as widely as the other from the true value.

The above remarks relate only to individual values of x. But the inequality $\pi(x) < \operatorname{li} x$ and Riemann's formula acquire some significance when considered from the point of view of averages, at any rate if the Riemann hypothesis is true. Thus (assuming the Riemann hypothesis in what follows) we have by Theorem 28, for all sufficiently large x,

$$(35) \qquad \left| \frac{1}{x} \int_0^x R(u)\,du \right| < x^{\frac{1}{2}} \sum_\rho \frac{1}{|\rho(\rho+1)|} + A < \frac{1}{10} x^{\frac{1}{2}}$$

[1] See Hardy's *Orders of Infinity* (Cambridge Tracts in Math. and Math. Physics, No. 12), Appendix III, Table 1 (1910); Appendix, Table 1 (1924).

[2] H, ii, 579–580.

[3] Lehmer 1, XIII–XVI; J. Glaisher 2, 84–88.

(the numerical inequality being an easy deduction from (25) and (27), p. 58); and a corresponding inequality for $Q(x)$ may be deduced. Using this in (9), we can show that

$$\int_2^x P(u)\,du < 0 \qquad (x > x_0),$$

so that $P(x)$ is 'negative on the average'. And we can see in the same way that the function $\operatorname{li} x - \frac{1}{2}\operatorname{li} x^{\frac{1}{2}}$ is 'on the average' a better approximation than $\operatorname{li} x$ to $\pi(x)$; but no importance can be attached to the later terms in Riemann's formula even by repeated averaging.

We see from (35) that the oscillations of $R(x)$ corresponding to the factor $\log\log\log x$ in Theorem 34 are smoothed out by the process of averaging. We may add that this must be due to the sparse distribution of abnormally large values of $|R(x)|$, and not merely to the cancelling of positive and negative contributions. For Cramér[1] has shown that (on the Riemann hypothesis)

$$\int_0^x \frac{R^2(u)}{u}\,du = O(x),$$

whence, by Schwarz's inequality,

$$\frac{1}{x}\int_0^x |R(u)|\,du = O(x^{\frac{1}{2}}).$$

The true orders of magnitude of $P(x)$, $Q(x)$ and $R(x)$ are not known, even on the Riemann hypothesis. The gap between the results of Theorem 30 on the one hand and those of Theorems 34 and 35 on the other represents one of the most important unsolved problems in the subject.

11. We conclude with some remarks on the distribution of the odd primes between the two arithmetical progressions $4n+1$ and $4n+3$.[2] We shall denote by $\pi^{(r)}(x)\,(r = 1, 3)$ the number of primes of the form $4n+r$ not exceeding x.

The analytical discussion of the problem is based on the function $L(s)$ defined (in the first instance for $\sigma > 0$) by the series

$$L(s) = \sum_1^\infty \frac{\chi(n)}{n^s},$$

where $\chi(n) = 0, 1, 0, -1$, according as $n \equiv 0, 1, 2, 3 \pmod{4}$. Since $\chi(n)/n^s$ is 'completely multiplicative' (p. 16), we have

$$L(s) = \prod_p \left(1 - \frac{\chi(p)}{p^s}\right)^{-1} \qquad (\sigma > 1).$$

The function $L(s)$ has properties very similar to those of $\zeta(s)$, with the

[1] Cramér 3. See also Cramér 4, 5; BC, 791–792; V, ii, 151–156.

[2] For an account of the general theory of which this is a special case (primes in arithmetical progressions of difference k) see H, i, 391–535; ii, 699–719: V, i, 79–96; ii, 3–47.

important difference that $L(s)$ is regular at all finite points, including the point $s=1$. The functions analogous to $\pi(x)$ and $\Pi(x)$ are

(36) $$\pi(x,\chi)=\sum_{p\leqslant x}\chi(p)=\pi^{(1)}(x)-\pi^{(3)}(x),\quad \Pi(x,\chi)=\sum_{p^m\leqslant x}\frac{\chi(p^m)}{m}.$$

By the use of the function $L(s)$ we can prove (i) by 'elementary' methods (see p. 39) that there exists an infinity of primes of each of the forms $4n+1$ and $4n+3$, and that $\Sigma 1/p$ diverges if taken over either of these sets of primes; and (ii) by 'transcendental' methods that

(37) $$\frac{\pi^{(1)}(x)}{\pi^{(3)}(x)}\to 1 \text{ as } x\to\infty.$$

Theorem (i) depends on the fact that $L(1)\neq 0$[1], and Theorem (ii) on the fact that $L(s)$ has no zeros on the line $\sigma=1$. By using the deeper properties of $L(s)$ we can prove (after Littlewood) that

$$\pi^{(1)}(x)-\pi^{(3)}(x)=O\left(xe^{-a\sqrt{\log x\log\log\log x}}\right);$$

and, if the analogue of the Riemann hypothesis is true for $L(s)$, still more precise results may be obtained.

Since $\chi(p^2)=+1$ for every odd p, we deduce from (36) that

$$\Pi(x,\chi)-\pi(x,\chi)=\tfrac{1}{2}\pi(x^{\frac{1}{2}})+O(x^{\frac{1}{3}}\log x)\sim\frac{x^{\frac{1}{2}}}{\log x},$$

so that

$$\pi(x,\chi)=\frac{x^{\frac{1}{2}}}{\log x}\left(-1+\Pi(x,\chi)\frac{\log x}{x^{\frac{1}{2}}}+o(1)\right).$$

Now, since $\Pi(x,\chi)$ (as the function 'naturally' associated with $\log L(s)$) may be expected to have its values fairly evenly distributed on either side of 0, this suggests that the values of $\pi(x,\chi)=\pi^{(1)}(x)-\pi^{(3)}(x)$ will be predominantly negative. And (as stated in the Introduction) empirical evidence points in the same direction, and indeed suggests that $\pi^{(1)}(x)<\pi^{(3)}(x)$ for all sufficiently large x. It can be proved, however, by the methods of §§ 7–9, that

$$\Pi(x,\chi)\frac{\log x}{x^{\frac{1}{2}}}=\Omega_{\pm}(\log\log\log x),$$

from which we conclude that, in fact, $\pi^{(1)}(x)-\pi^{(3)}(x)$ changes sign infinitely often as x increases to infinity. But, if the analogue of the Riemann hypothesis is true, there are various senses in which the primes of the form $4n+3$ are 'on the average' more numerous than those of the form $4n+1$.[2]

[1] The relation $L(1)\neq 0$ is, of course, trivial in the special case under discussion ($k=4$). But the corresponding relations in the general case, though still 'elementary', constitute one of the main difficulties of the problem. The existence of an infinity of primes in each of the arithmetical progressions $4n+1$ and $4n+3$ can be proved by a modification of Euclid's method (pp. 1–2), the case $4n+1$ being the more difficult. The same method is applicable to some other special values of k, but has not been extended to the general case.

[2] Cf. Hardy and Littlewood 2, 141–151; Landau 6, 7.

BIBLIOGRAPHY

[The following list makes no attempt at completeness. For a full account of the subject the reader is referred to the two monumental works of Landau (**H** and **V**), and for an admirable summary to the encyclopædia article of Bohr and Cramér (**BC**); detailed references to these works have not been given in the course of the tract. **H** contains an exhaustive bibliography up to 1909, and **BC** full references up to 1922. For literature relating to the pure theory of the zeta-function we refer the reader to Titchmarsh's tract (**T**).]

[The dates refer in general to the publication of the volume in which a paper appears. Where two dates are given the first is that of reading or presenting the paper, the second that of publication of the volume.]

H. E. LANDAU, *Handbuch der Lehre von der Verteilung der Primzahlen* (Leipzig, Teubner, 1909).

BHM. P. BACHMANN—J. HADAMARD—E. MAILLET, Propositions transcendantes de la théorie des nombres, *Encyclopédie des sciences mathématiques*, I 17 (1910), 215–387.

HR. G. H. HARDY—M. RIESZ, *The general theory of Dirichlet's series* (Cambridge Tracts in Math. and Math. Physics, No. 18, 1915).

BC. H. BOHR—H. CRAMÉR, Die neuere Entwicklung der analytischen Zahlentheorie, *Enzyklopädie der mathematischen Wissenschaften*, II C 8 (1922), 722–849.

V. E. LANDAU, *Vorlesungen über Zahlentheorie* (Leipzig, Hirzel, 1927).

T. E. C. TITCHMARSH, *The zeta-function of Riemann* (Cambridge Tracts in Math. and Math. Physics, No. 26, 1930).

R. J. Backlund

1. Sur les zéros de la fonction $\zeta(s)$ de Riemann, *Comptes rendus*, 158 (1914), 1979–1981.
2. Über die Nullstellen der Riemannschen Zetafunktion, *Acta Math.*, 41 (1918), 345–375.

R. Breusch

1. Zur Verallgemeinerung des Bertrandschen Postulates, dass zwischen x und $2x$ stets Primzahlen liegen, *Math. Zeitschrift*, 34 (1932), 505–526.

V. Brun

1. Le crible d'Eratosthène et le théorème de Goldbach, *Skrifter utgit av Videnskapsselskapet i Kristiania, mat.-naturv. Kl.* $(1920)_1$: 3.

P. L. Chebyshev (= Tschebyscheff, etc.)

1. Теорія Сравненій (St Petersburg, 1849), Appendix III.
2. Sur la fonction qui détermine la totalité des nombres premiers inférieurs à une limite donnée; **a**, *Mémoires présentés à l'Académie Impériale des sciences de St.-Pétersbourg par divers savants*, 6 (1848; 1851), 141–157; **b**, *Journal de math.* (1), 17 (1852), 341–365. [*Œuvres*, i, 27–48.]
3. Mémoire sur les nombres premiers; **a**, *Mémoires présentés à l'Académie Impériale des sciences de St.-Pétersbourg par divers savants*, 7 (1850; 1854) 15–33; **b**, *Journal de math.* (1), 17 (1852), 366–390. [*Œuvres*, i, 49–70.]

H. Cramér

1. Über die Herleitung der Riemannschen Primzahlformel, *Arkiv för Mat., Astr. och Fys.*, 13 (1918): 24.
2. Studien über die Nullstellen der Riemannschen Zetafunktion, *Math. Zeitschrift*, 4 (1919), 104–130.
3. Some theorems concerning prime numbers, *Arkiv för Mat., Astr. och Fys.*, 15 (1921): 5.
4. Sur un problème de M. Phragmén, *Arkiv för Mat., Astr. och Fys.*, 16 (1922): 27.
5. Ein Mittelwertsatz in der Primzahltheorie, *Math. Zeitschrift*, 12 (1922), 147–153.

L. Euler

1. Variae observationes circa series infinitas, *Commentarii Academiae Scientiarum Imperialis Petropolitanae*, 9 (1737; 1744), 160–188. [*Opera omnia* (1), 14, 216–244.]
2. *Introductio in analysin infinitorum*, Vol. 1 (Lausanne, Bousquet, 1748). [*Opera omnia* (1), 8.]

C. F. Gauss

1. *Werke*; **a**, 1st edition (Göttingen, 1863); **b**, 2nd edition (Göttingen, 1876).

J. Glaisher

1. *Factor table for the fourth million* (London, Taylor and Francis, 1879).
2. *Factor table for the sixth million* (London, Taylor and Francis, 1883).

J. W. L. Glaisher

1. Separate enumerations of primes of the form $4n + 1$ and of the form $4n + 3$, *Proc. Royal Soc.*, 29 (1879), 192–197.

J. Hadamard

1. Étude sur les propriétés des fonctions entières et en particulier d'une fonction considérée par Riemann, *Journal de math.* (4), 9 (1893), 171–215.
2. Sur la distribution des zéros de la fonction $\zeta(s)$ et ses conséquences arithmétiques, *Bulletin de la Soc. math. de France*, 24 (1896), 199–220.

G. H. Hardy

1. A new proof of the functional equation for the zeta-function, *Matematisk Tidsskrift* B (1922), 71–73.
2. Note on a theorem of Mertens, *Journal London Math. Soc.*, 2 (1927), 70–72.

G. H. Hardy and J. E. Littlewood

1. New proofs of the prime-number theorem and similar theorems, *Quarterly Journal of Math.*, 46 (1915), 215–219.
2. Contributions to the theory of the Riemann zetafunction and the theory of the distribution of primes, *Acta Math.*, 41 (1918), 119–196.
3. On a Tauberian theorem for Lambert's series, and some fundamental theorems in the analytical theory of numbers, *Proc. London Math. Soc.* (2), 19 (1921), 21–29.
4. Some problems of 'Partitio numerorum'; III. On the expression of a number as a sum of primes, *Acta Math.*, 44 (1922), 1–70.
5. Some problems of 'Partitio numerorum' (V): A further contribution to the study of Goldbach's problem, *Proc. London Math. Soc.* (2), 22 (1923), 46–56.

C. J. Hargreave

1. Analytical researches concerning numbers, *Philosophical Magazine* (3), 35 (1849), 36–53.
2. On the law of prime numbers, *Philosophical Magazine* (4), 8 (1854), 114–122.

G. Hoheisel

1. Primzahlprobleme in der Analysis, *Sitzungsberichte d. Preuss. Akad. d. Wissens., phys.-math. Kl.* (Berlin, 1930), 580–588.

E. Holmgren

1. Om primtalens fördelning, *Öfversigt af Kongl. Vetenskaps-Akademiens Förhandlingar*, 59 (1902), 221–225.

S. Ikehara

1. An extension of Landau's theorem in the analytical theory of numbers, *Journal of Math. and Phys., Massachusetts Inst. of Technology*, 10 (1931), 1–12.

A. E. Ingham

1. Note on Riemann's ζ-function and Dirichlet's L-functions, *Journal London Math. Soc.*, 5 (1930), 107–112.

J. Karamata

1. Über die Hardy-Littlewoodschen Umkehrungen des Abelschen Stetigkeitssatzes, *Math. Zeitschrift*, 32 (1930), 319–320.

H. von Koch

1. Sur la distribution des nombres premiers, *Acta Math.*, 24 (1901), 159–182.

E. Landau

1. Über einen Satz von Tschebyschef, *Math. Annalen*, 61 (1905), 527–550.

2. Euler und die Funktionalgleichung der Riemannschen Zetafunktion, *Bibliotheca Mathematica* (3), 7 (1906), 69–79.

3. Über die Äquivalenz zweier Hauptsätze der analytischen Zahlentheorie, *Sitzungsberichte d. Akad. d. Wissens. in Wien, math.-naturw. Kl.*, 120, Abt. 2*a* (1911), 973–988.

4. Über einige Summen, die von den Nullstellen der Riemann'schen Zetafunktion abhängen, *Acta Math.*, 35 (1912), 271–294.

5. Über die Nullstellen der Zetafunktion, *Math. Annalen*, 71 (1912), 548–564.

6, 7. Über einige ältere Vermutungen und Behauptungen in der Primzahltheorie, *Math. Zeitschrift*, 1 (1918), 1–24; (zweite Abhandlung), 213–219.

8. Über die ζ-Funktion und die L-Funktionen, *Math. Zeitschrift*, 20 (1924), 105–125.

9. Über die Zetafunktion und die Hadamardsche Theorie der ganzen Funktionen, *Math. Zeitschrift*, 26 (1927), 170–175.

10. Die Goldbachsche Vermutung und der Schnirelmannsche Satz, *Nachrichten v. d. Gesellschaft der Wissens. zu Göttingen, math.-phys. Kl.* (1930), 255–276.

A. M. Legendre

1. *Essai sur la théorie des nombres*; a, 1st edition (Paris, Duprat, 1798). b, 2nd edition (Paris, Courcier, 1808).

2. *Théorie des nombres* (Paris, Didot, 1830; 3rd edition of 1).

D. N. Lehmer

1. *List of prime numbers from* 1 *to* 10,006,721 (Carnegie Institution of Washington, Publication No. 165, Washington, D.C., 1914).

J. E. Littlewood (*see also* **G. H. Hardy**)

1. Sur la distribution des nombres premiers, *Comptes Rendus*, 158 (1914), 1869–1872.

2. Researches in the theory of the Riemann ζ-function, *Proc. London Math. Soc.* (2) 20 (1922), xxii–xxviii (*Records*, Feb. 10, 1921).

3. Two notes on the Riemann zeta-function, *Proc. Cambridge Philos. Soc.*, 22 (1924), 234–242.

4. Mathematical notes: 3; On a theorem concerning the distribution of prime numbers, *Journal London Math. Soc.*, 2 (1927), 41–45.

H. von Mangoldt

1. Auszug aus einer Arbeit unter dem Titel: Zu Riemann's Abhandlung 'Über die Anzahl der Primzahlen unter einer gegebenen Grösse', *Sitzungsberichte d. Preuss. Akad. d. Wissens.* (Berlin, 1894), 883–896.

2. Zu Riemann's Abhandlung 'Ueber die Anzahl der Primzahlen unter einer gegebenen Grösse', *Journal für die r. u. a. Math.*, 114 (1895), 255–305.

3. Beweis der Gleichung $\sum_{k=1}^{\infty} \frac{\mu(k)}{k} = 0$, *Sitzungsberichte d. Preuss. Akad. d. Wissens.* (Berlin, 1897), 835–852.

4. Zur Verteilung der Nullstellen der Riemannschen Funktion $\xi(t)$, *Math. Annalen*, 60 (1905), 1–19.

F. Mertens

1. Ein Beitrag zur analytischen Zahlentheorie, *Journal für die r. u. a. Math.*, 78 (1874), 46–62.

2. Über eine Eigenschaft der Riemann'schen ζ-Function, *Sitzungsberichte d. Akad. d. Wissens. in Wien, math.-naturw. Cl.*, 107, Abt. 2*a* (1898), 1429–1434.

L. J. Mordell

1. Some applications of Fourier series in the analytic theory of numbers, *Proc. Cambridge Philos. Soc.*, 24 (1928), 585–596.

E. Phragmén

1. Sur le logarithme intégral et la fonction $f(x)$ de Riemann, *Öfversigt af Kongl. Vetenskaps-Akademiens Förhandlingar*, 48 (1891), 599–616.

2. Sur une loi de symétrie relative à certaines formules asymptotiques, *Öfversigt af Kongl. Vetenskaps-Akademiens Förhandlingar*, 58 (1901), 189–202.

G. Pólya

1. Über das Vorzeichen des Restgliedes im Primzahlsatz, *Nachrichten v. d. Gesellschaft der Wissens. zu Göttingen, math.-phys. Kl.* (1930), 19–27.

H. Rademacher

1. Beiträge zur Viggo Brunschen Methode in der Zahlentheorie, *Abhandlungen aus d. math. Seminar d. Hamburgischen Univ.*, 3 (1924), 12–30.

B. Riemann

1. Ueber die Anzahl der Primzahlen unter einer gegebenen Grösse, *Monatsberichte d. Preuss. Akad. d. Wissens.* (Berlin: 1859; 1860), 671–680. [*Gesammelte mathematische Werke* (1st ed., 1876), 136–144, (2nd ed., 1892), 145–155; *Œuvres mathématiques* (1898), 165–176.]

E. Schmidt

1. Über die Anzahl der Primzahlen unter gegebener Grenze, *Math. Annalen*, 57 (1903), 195–204.

L. Schnirelmann

1. Об аддитивных свойствах чисел, *Известия Донского Политехнического Института* (Новочеркасск), 14 (1930), 3–28.

I. Schur

1, 2. Einige Sätze über Primzahlen mit Anwendungen auf Irreduzibilitätsfragen. I, *Sitzungsberichte d. Preuss. Akad. d. Wissens.* (Berlin, 1929), 125–136; II, *ibid.* (1929), 370–391.

C. L. Siegel

1. Über Riemanns Nachlass zur analytischen Zahlentheorie, *Quellen und Studien zur Geschichte der Math., Astr. und Phys.*, Abt. B, 2 (1932), 45–80.

J. J. Sylvester

1. On Tchebycheff's theory of the totality of prime numbers comprised within given limits, *American Journal of Math.*, 4 (1881), 230–247.

2, 3. On arithmetical series, *Messenger of Math.* (2), 21 (1892), 1–19; 87–120.

P. L. Tschebyscheff, etc. *See* **P. L. Chebyshev**

C.-J. de la Vallée Poussin

1. Recherches analytiques sur la théorie des nombres; Première partie: La fonction $\zeta(s)$ de Riemann et les nombres premiers en général, *Annales de la Soc. scientifique de Bruxelles*, 20_2 (1896), 183–256.
2. Sur la fonction $\zeta(s)$ de Riemann et le nombre des nombres premiers inférieurs à une limite donnée, *Mémoires couronnés de l'Acad. roy. des Sciences...de Belgique*, 59 (1899–1900): 1.

N. Wiener

1. A new method in Tauberian theorems, *Journal of Math. and Phys., Massachusetts Inst. of Technology*, 7 (1927–1928), 161–184.
2. Tauberian theorems, *Annals of Math.* (2), 33 (1932), 1–100.

Printed in the United States
By Bookmasters